50% OFF Online ACT Prep Course!

Dear Customer,

We consider it an honor and a privilege that you chose our ACT Study Guide. As a way of showing our appreciation and to help us better serve you, we have partnered with Mometrix Test Preparation to offer you **50% off their online ACT Course**. Many ACT courses are needlessly expensive and don't deliver enough value. With their course, you get access to the best ACT prep material, and you only pay half price.

Mometrix has structured their online course to perfectly complement your printed study guide. The ACT Online Course contains **in-depth lessons** that cover all the most important topics, **180+ video reviews** that explain difficult concepts, over **1,500 practice questions** to ensure you feel prepared, and more than **500 digital flashcards**, so you can study while you're on the go.

Online ACT Prep Course

Topics Covered:

- English
  - Organization, Unity, and Cohesion
  - Punctuation and Grammar
- Math
  - Numbers and Operations
  - Algebra and Geometry
- Reading
  - Key Ideas and Details
  - Purpose, Position, and Point of View
- Science
  - Evaluation of Models
  - Interpretation of Data
- Writing

Course Features:

- ACT Study Guide
  - Get content that complements our best-selling study guide.
- Full-Length Practice Tests
  - With over 1,500 practice questions, you can test yourself again and again.
- Mobile Friendly
  - If you need to study on the go, the course is easily accessible from your mobile device.
- ACT Flashcards
  - Their course includes a flashcards mode with over 500 content cards for you to study.

To receive this discount, visit their website: mometrix.com/university/act or simply scan this QR code with your smartphone. At the checkout page, enter the discount code: **APEXACT50**

If you have any questions or concerns, please contact them at universityhelp@mometrix.com.

SCAN HERE

FREE

Free Study Tips DVD

In addition to the tips and content in this guide, we have created a FREE DVD with helpful study tips to further assist your exam preparation. **This FREE Study Tips DVD provides you with top-notch tips to conquer your exam and reach your goals.**

Our simple request in exchange for the strategy-packed DVD is that you email us your feedback about our study guide. We would love to hear what you thought about the guide, and we welcome any and all feedback—positive, negative, or neutral. It is our #1 goal to provide you with top-quality products and customer service.

To receive your **FREE Study Tips DVD**, email freedvd@apexprep.com. Please put "FREE DVD" in the subject line and put the following in the email:

a. The name of the study guide you purchased.

b. Your rating of the study guide on a scale of 1-5, with 5 being the highest score.

c. Any thoughts or feedback about your study guide.

d. Your first and last name and your mailing address, so we know where to send your free DVD!

Thank you!

ACT Math Prep Book 2021 and 2022

Study Guide with 3 Practice Tests [3rd Edition]

Matthew Lanni

Written and edited by APEX Publishing.

ISBN 13: 9781637754306
ISBN 10: 1637754302

For additional information or for bulk orders, contact info@apexprep.com.

Table of Contents

Test Taking Strategies

1. Reading the Whole Question

A popular assumption in Western culture is the idea that we don't have enough time for anything. We speed while driving to work, we want to read an assignment for class as quickly as possible, or we want the line in the supermarket to dwindle faster. However, speeding through such events robs us from being able to thoroughly appreciate and understand what's happening around us. While taking a timed test, the feeling one might have while reading a question is to find the correct answer as quickly as possible. Although pace is important, don't let it deter you from reading the whole question. Test writers know how to subtly change a test question toward the end in various ways, such as adding a negative or changing focus. If the question has a passage, carefully read the whole passage as well before moving on to the questions. This will help you process the information in the passage rather than worrying about the questions you've just read and where to find them. A thorough understanding of the passage or question is an important way for test takers to be able to succeed on an exam.

2. Examining Every Answer Choice

Let's say we're at the market buying apples. The first apple we see on top of the heap may *look* like the best apple, but if we turn it over we can see bruising on the skin. We must examine several apples before deciding which apple is the best. Finding the correct answer choice is like finding the best apple. Although it's tempting to choose an answer that seems correct at first without reading the others, it's important to read each answer choice thoroughly before making a final decision on the answer. The aim of a test writer might be to get as close as possible to the correct answer, so watch out for subtle words that may indicate an answer is incorrect. Once the correct answer choice is selected, read the question again and the answer in response to make sure all your bases are covered.

3. Eliminating Wrong Answer Choices

Sometimes we become paralyzed when we are confronted with too many choices. Which frozen yogurt flavor is the tastiest? Which pair of shoes look the best with this outfit? What type of car will fill my needs as a consumer? If you are unsure of which answer would be the best to choose, it may help to use process of elimination. We use "filtering" all the time on sites such as eBay® or Craigslist® to eliminate the ads that are not right for us. We can do the same thing on an exam. Process of elimination is crossing out the answer choices we know for sure are wrong and leaving the ones that might be correct. It may help to cover up the incorrect answer choice. Covering incorrect choices is a psychological act that alleviates stress due to the brain being exposed to a smaller amount of information. Choosing between two answer choices is much easier than choosing between all of them, and you have a better chance of selecting the correct answer if you have less to focus on.

4. Sticking to the World of the Question

When we are attempting to answer questions, our minds will often wander away from the question and what it is asking. We begin to see answer choices that are true in the real world instead of true in the world of the question. It may be helpful to think of each test question as its own little world. This world may be different from ours. This world may know as a truth that the chicken came before the egg or may assert that two plus two equals five. Remember that, no matter what hypothetical nonsense may be in the question, assume it to be true. If the question states that the chicken came before the egg, then choose your answer based on that truth. Sticking to the world of the question means placing all of our biases and assumptions aside and relying on the question to guide us to the correct answer. If we are simply looking for answers that are correct based on our own judgment, then we may choose incorrectly. Remember an answer that is true does not necessarily answer the question.

5. Key Words

If you come across a complex test question that you have to read over and over again, try pulling out some key words from the question in order to understand what exactly it is asking. Key words may be words that surround the question, such as *main idea, analogous, parallel, resembles, structured,* or *defines*. The question may be asking for the main idea, or it may be asking you to define something. Deconstructing the sentence may also be helpful in making the question simpler before trying to answer it. This means taking the sentence apart and obtaining meaning in pieces, or separating the question from the foundation of the question. For example, let's look at this question:

> Given the author's description of the content of paleontology in the first paragraph, which of the following is most parallel to what it taught?

The question asks which one of the answers most *parallels* the following information: The *description* of paleontology in the first paragraph. The first step would be to see *how* paleontology is described in the first paragraph. Then, we would find an answer choice that parallels that description. The question seems complex at first, but after we deconstruct it, the answer becomes much more attainable.

6. Subtle Negatives

Negative words in question stems will be words such as *not, but, neither,* or *except*. Test writers often use these words in order to trick unsuspecting test takers into selecting the wrong answer—or, at least, to test their reading comprehension of the question. Many exams will feature the negative words in all caps (*which of the following is NOT an example*), but some questions will add the negative word seamlessly into the sentence. The following is an example of a subtle negative used in a question stem:

> According to the passage, which of the following is *not* considered to be an example of paleontology?

If we rush through the exam, we might skip that tiny word, *not*, inside the question, and choose an answer that is opposite of the correct choice. Again, it's important to read the question fully, and double check for any words that may negate the statement in any way.

7. Spotting the Hedges

The word "hedging" refers to language that remains vague or avoids absolute terminology. Absolute terminology consists of words like *always, never, all, every, just, only, none,* and *must*. Hedging refers to words like *seem, tend, might, most, some, sometimes, perhaps, possibly, probability,* and *often*. In some cases, we want to choose answer choices that use hedging and avoid answer choices that use absolute terminology. It's important to pay attention to what subject you are on and adjust your response accordingly.

8. Restating to Understand

Every now and then we come across questions that we don't understand. The language may be too complex, or the question is structured in a way that is meant to confuse the test taker. When you come across a question like this, it may be worth your time to rewrite or restate the question in your own words in order to understand it better. For example, let's look at the following complicated question:

> Which of the following words, if substituted for the word *parochial* in the first paragraph, would LEAST change the meaning of the sentence?

Let's restate the question in order to understand it better. We know that they want the word *parochial* replaced. We also know that this new word would "least" or "not" change the meaning of the sentence. Now let's try the sentence again:

> Which word could we replace with *parochial,* and it would not change the meaning?

Restating it this way, we see that the question is asking for a synonym. Now, let's restate the question so we can answer it better:

> Which word is a synonym for the word *parochial*?

Before we even look at the answer choices, we have a simpler, restated version of a complicated question.

9. Predicting the Answer

After you read the question, try predicting the answer *before* reading the answer choices. By formulating an answer in your mind, you will be less likely to be distracted by any wrong answer choices. Using predictions will also help you feel more confident in the answer choice you select. Once you've chosen your answer, go back and reread the question-and-answer choices to make sure you have the best fit. If you have no idea what the answer may be for a particular question, forego using this strategy.

10. Avoiding Patterns

One popular myth in grade school relating to standardized testing is that test writers will often put multiple-choice answers in patterns. A runoff example of this kind of thinking is that the most common answer choice is "C," with "B" following close behind. Or, some will advocate certain made-up word patterns that simply do not exist. Test writers do not arrange their correct answer choices in any kind of pattern; their choices are randomized. There may even be times where the correct answer choice will be the same letter for two or three questions in a row, but we have no way of knowing when or if this might happen. Instead of trying to figure out what choice the test writer probably set as being correct, focus on what the *best answer choice* would be out of the answers you are presented with. Use the tips above, general knowledge, and reading comprehension skills in order to best answer the question, rather than looking for patterns that do not exist.

FREE DVD OFFER

Achieving a high score on your exam depends not only on understanding the content, but also on understanding how to apply your knowledge and your command of test taking strategies. **Because your success is our primary goal, we offer a FREE Study Tips DVD. It provides top-notch test taking strategies to help you optimize your testing experience.**

Our simple request in exchange for the strategy-packed DVD is that you email us your feedback about our study guide.

To receive your **FREE Study Tips DVD**, email freedvd@apexprep.com. Please put "FREE DVD" in the subject line and put the following in the email:

a. The name of the study guide you purchased.

b. Your rating of the study guide on a scale of 1-5, with 5 being the highest score.

c. Any thoughts or feedback about your study guide.

d. Your first and last name and your mailing address, so we know where to send your free DVD!

Introduction to the ACT

Function of the Test

The ACT is one of the two national standardized college entrance examinations, with the SAT serving as the other option. Most hopeful college-bound students take the ACT, SAT, or both. For admissions purposes, every four-year college and university in the United States accepts ACT scores, and some schools require it. More so than the SAT, which primarily serves as an aptitude test, the ACT is often used for course placement purposes because it measures academic achievement on content addressed in high school classes. Twelve states also require that all high school juniors in the state take the ACT, and eight additional states have counties that require the exam.

Most ACT test takers are prospective college students who are currently in their junior or senior year of high school. More than 2 million students graduating in the class of 2017 took the ACT.

Test Administration

The ACT is offered on seven dates in the U.S. and Canada, and on six dates internationally throughout the year. The exam is usually administered at high schools or colleges, but other locations may be offered. The registration fee includes the cost to submit score reports to four colleges, but for an additional fee, students can send scores to additional institutions. There is a separate registration fee incurred for the optional writing section. Some high schools cover the fees for their students, so prospective test takers are advised to contact the guidance counselor at their school.

Test takers can retake the ACT every time the test is offered, up to a maximum of twelve times. However, different colleges and universities sometimes have limits on the number of retakes they will consider. Beginning September 2020, test takers wishing to take one or several of the five sections of the ACT can do so without needing to retake all sections. This can be advantageous for test takers who do not want to risk lowering their scores on sections over which they previously performed well.

Reasonable accommodations will be provided to test takers with appropriate documentation for a variety of disabilities.

Test Format

Test takers are given a total of 175 minutes to complete the 215 multiple-choice questions in four subject subtests (English, Mathematics, Reading, and Science) of the ACT. It also has an optional Writing Test, which involves writing an essay, which takes an additional forty minutes. Some colleges and universities require the essay for admission.

The English Test consists of 75 questions that address the production of writing, knowledge of language, and conventions of standard English. The 60-question Mathematics Test involves number sense, algebra, functions, geometry, statistics and probability, and modeling. Calculators that meet certain calculator requirements are permitted. The Reading Test contains four written passages, with ten questions per passage addressing comprehension skills, the ability to make inferences and draw conclusions, and apply and integrate knowledge. The Science Test contains 40 questions that require interpreting data, understanding scientific investigations, and evaluating models and results.

The Writing Test is always given at the end of the exam so that test takers opting not to take it may leave after completing the other four subtests. This section consists of one essay in which students must analyze three different perspectives on a broad social issue and reconcile them in a cohesive essay. The following chart provides the breakdown of the sections of the ACT:

Subtest	Length	Number of Questions
English	45 minutes	75
Mathematics	60 minutes	60
Reading	35 minutes	40
Science	35 minutes	40
Writing (optional)	40 minutes	1 essay

Scoring

Score reports are typically available two weeks after the date of administration. Because there is no penalty for incorrect answers, test takers are encouraged to answer every question, even if they have to guess. For each of the four required subtests, test takers receive a score between 1 and 36. These scores are then averaged together to yield a Composite Score, which is the primary score reported as an "ACT score." The most prestigious colleges and universities are typically looking for Composite Scores greater than 30 in order to consider an applicant for admissions. Other selective schools typically expect candidates to have scores just under 30. Average institutions are more likely to set the bar lower (perhaps in the low 20s), while community colleges usually accept students with scores in the high teens. In 2016 and 2017, the mean Composite Score among all test takers (including those not applying to college) was 20.9.

The Writing Test is scored on a scale that ranges from 2 to 12 scale. In 2016 and 2017, the mean score was 6.7.

Prior to September 2020, test takers who chose to retake the exam were required to retake all sections of the exam. Then, their composite score would simply be reflective of their performance on the subsections of the retake. Scores from individual subsections from different test attempts could not be combined. As of September 2020, a new "superscore" can be generated. This score is a conglomerate of the test taker's highest scores on subsections from any of their attempts at the test.

Mathematics Test

Preparing for Higher Mathematics

Number & Quantity

Real and Complex Number Systems

Whole numbers are the numbers 0, 1, 2, 3, ... Examples of other whole numbers would be 413 and 8,431. Notice that numbers such as 4.13 and $\frac{1}{4}$ are not included in whole numbers. **Counting numbers**, also known as **natural numbers**, consist of all whole numbers except for the zero. In set notation, the natural numbers are the set $\{1, 2, 3, \ldots\}$. The entire set of whole numbers and negative versions of those same numbers comprise the set of numbers known as **integers.** Therefore, in set notation, the integers are $\{\ldots, -3, -2, -1, 0, 1, 2,3, \ldots\}$. Examples of other integers are −4,981 and 90,131. A number line is a great way to visualize the integers. Integers are labeled on the following number line:

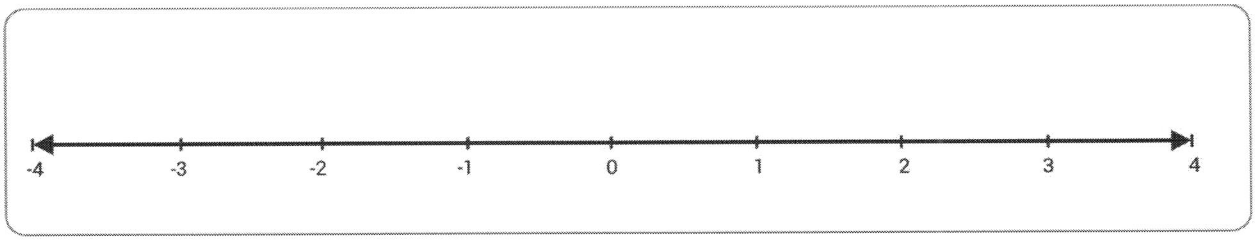

The arrows on the right- and left-hand sides of the number line show that the line continues indefinitely in both directions.

Fractions also exist on the number line and are considered parts of a whole. For example, if an entire pie is cut into two pieces, each piece is half of the pie, or $\frac{1}{2}$. The top number in any fraction, known as the **numerator,** defines how many parts there are. The bottom number, known as the **denominator**, states how many pieces the whole is divided into. Fractions can also be negative or written in their corresponding decimal form.

A **decimal** is a number that uses a decimal point and numbers to the right of the decimal point representing the part of the number that is less than 1. For example, 3.5 is a decimal and is equivalent to the fraction $\frac{7}{2}$ or the mixed number $3\frac{1}{2}$. The decimal is found by dividing 2 into 7. Other examples of fractions are $\frac{2}{7}, \frac{-3}{14},$ and $\frac{14}{27}$.

Any number that can be expressed as a fraction is known as a **rational number.** Basically, if a and b are any integers and $b \neq 0$, then $\frac{a}{b}$ is a rational number. Any integer can be written as a fraction where the denominator is 1; therefore, the set of rational numbers consist of all fractions and all integers.

Any number that is not rational is known as an **irrational number.** Consider the number $\pi = 3.141592654$ .... The decimal portion of that number extends indefinitely. In that situation, the number can never be written as a fraction. Another example of an irrational number is $\sqrt{2} = 1.414213662$ .... Again, this number cannot be written as a ratio of two integers.

Together, the set of all rational and irrational numbers makes up the **real numbers.** The number line contains all real numbers. To graph a number other than an integer on a number line, the number will need to be plotted between two integers. For example, 3.5 would be plotted halfway between 3 and 4.

Even numbers are integers that are divisible by 2. For example, 6, 100, 0, and −200 are all even numbers. **Odd numbers** are integers that are not divisible by 2. If an odd number is divided by 2, the result is a fraction. For example, −5, 11, and −121 are odd numbers.

Prime numbers consist of natural numbers greater than 1 that are not divisible by any other natural numbers other than themselves and 1. For example, 3, 5, and 7 are prime numbers. If a natural number is not prime, it is known as a **composite number**. 8 is a composite number because it is divisible by both 2 and 4, which are natural numbers other than itself and 1.

The **absolute value** of any real number is the distance from that number to 0 on the number line. The absolute value of a number can never be negative. For example, the absolute value of both 8 and −8 is 8 because they are both 8 units away from 0 on the number line. This is written as $|8| = |-8| = 8$.

Performing Arithmetic Operations with Rational Numbers

The four basic operations include addition, subtraction, multiplication, and division. The result of addition is a **sum**, the result of subtraction is a **difference**, the result of multiplication is a **product**, and the result of division is a **quotient**. Each type of operation can be used when working with rational numbers; however, the basic operations need to be understood first while using simpler numbers before working with fractions and decimals.

These operations should first be learned using whole numbers. Addition needs to be done column by column. To add two whole numbers, add the ones column first, then the tens columns, then the hundreds, etc. If the sum of any column is greater than 9, a one must be carried over to the next column. For example, the following is the result of 482 + 924:

$$\begin{array}{r} {}^{1} \\ 482 \\ +924 \\ \hline 1406 \end{array}$$

Notice that the sum of the tens column was 10, so a one was carried over to the hundreds column. Subtraction is also performed column by column. Subtraction is performed in the ones column first, then the tens, etc. If the number on top is less than the number below, a one must be borrowed from the column to the left. For example, the following is the result of 5,424 – 756:

$$\begin{array}{cccc} 4 & 13 & 11 & 14 \\ \cancel{5} & \cancel{4} & \cancel{2} & \cancel{4} \\ - & 7 & 5 & 6 \\ \hline 4 & 6 & 6 & 8 \end{array}$$

Notice that a one is borrowed from the tens, hundreds, and thousands place. After subtraction, the answer can be checked through addition. A check of this problem would be to show that 756 + 4,668 = 5,424.

In multiplication, the number on top is known as the multiplicand, and the number below is the multiplier. Complete the problem by multiplying the multiplicand by each digit of the multiplier. Make sure to place the ones value of each result under the multiplying digit in the multiplier. The final product is found by adding each partial product. The following example shows the process of multiplying 46 times 37:

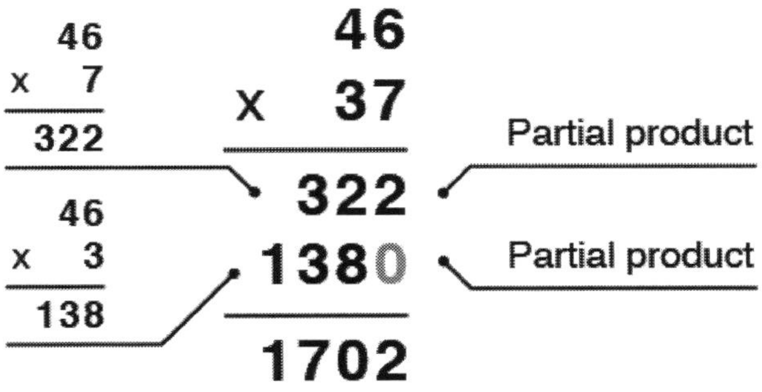

Finally, division can be performed using long division. When dividing, the first number is known as the **dividend,** and the second is the **divisor.** For example, with $a \div b = c$, a is the dividend, b is the divisor, and c is the quotient. For long division, place the dividend within the division bar and the divisor on the outside. For example, with $8{,}764 \div 4$, refer to the first problem in the diagram below. The first digit, 8, is divisible by 4 two times. Therefore, 2 goes above the division bar over the 8. Then, multiply 4 times 2 to get 8, and that product goes below the 8. Subtract to get 0, and then carry down the second digit, 7. Continue the same steps. $7 \div 4 = 1\text{ R}3$, so 1 is written above the 7. Multiply 4 times 1 to get 4, and write it below the 7. Subtract to get 3, and carry the 6 down next to the 3. Continuing this process for the next two digits results in a 9 and a 1. The final subtraction results in a 0, which means that 8,764 is evenly divisible by 4 with no remaining numbers.

The second example shows that:

$$4{,}536 \div 216 = 21$$

The steps are a little different because 216 cannot be contained in 4 or 5, so the first step is placing a 2 above the 3 because there are 2 216's in 453. Finally, the third example shows that:

$$546 \div 31 = 17\text{ R}19$$

The 19 is a remainder. Notice that the final subtraction does not result in a 0, which means that 546 is not divisible by 31.

The remainder can also be written as a fraction over the divisor to say that:

$$546 \div 31 = 17\frac{19}{31}$$

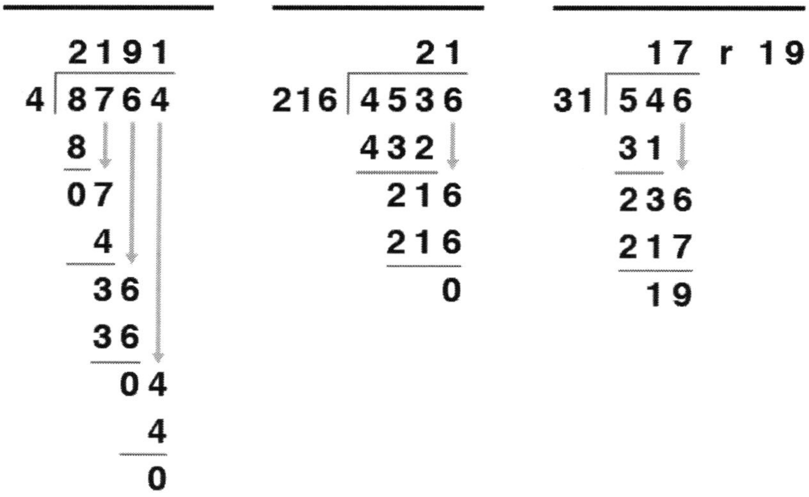

A remainder can have meaning in a division problem with real-world application. For example, consider the third example,

$$546 \div 31 = 17 \text{ R}19$$

Let's say that we had \$546 to spend on calculators that cost \$31 each, and we wanted to know how many we could buy. The division problem would answer this question. The result states that 17 calculators could be purchased, with \$19 left over. Notice that the remainder will never be greater than or equal to the divisor.

Once the operations are understood with whole numbers, they can be used with negative numbers. There are many rules surrounding operations with negative numbers. First, consider addition with integers. The sum of two numbers can first be shown using a number line. For example, to add $-5 + (-6)$, plot the point -5 on the number line. Adding a negative number is the same as subtracting, so move 6 units to the left. This process results in landing on -11 on the number line, which is the sum of -5 and -6. If adding a positive number, move to the right.

While visualizing this process using a number line is useful for understanding, it is more efficient to learn the rules of operations. When adding two numbers with the same sign, add the absolute values of both numbers, and use the common sign of both numbers as the sign of the sum. For example, to add $-5 + (-6)$, add their absolute values:

$$5 + 6 = 11$$

Then, introduce a negative number because both addends are negative. The result is -11.

To add two integers with unlike signs, subtract the lesser absolute value from the greater absolute value, and apply the sign of the number with the greater absolute value to the result. For example, the sum

$-7+4$ can be computed by finding the difference $7-4=3$ and then applying a negative because the value with the larger absolute value is negative. The result is −3. Similarly, the sum $-4+7$ can be found by computing the same difference but leaving it as a positive result because the addend with the larger absolute value is positive. Also, recall that any number plus 0 equals that number. This is known as the **Addition Property of 0.**

Subtracting two integers with opposite signs can be computed by changing to addition to avoid confusion. The rule is to add the first number to the opposite of the second number. The opposite of a number is the number with the same value on the other side of 0 on the number line. For example, −2 and 2 are opposites. Consider $4-8$. Change this to adding the opposite as follows: $4+(-8)$. Then, follow the rules of addition of integers to obtain −4. Secondly, consider $-8-(-2)$. Change this problem to adding the opposite as $-8+2$, which equals −6. Notice that subtracting a negative number functions the same as adding a positive number.

Multiplication and division of integers are actually less confusing than addition and subtraction because the rules are simpler to understand. If two factors in a multiplication problem have the same sign, the result is positive. If one factor is positive and one factor is negative, the result, known as the **product,** is negative. For example,

$$(-9)(-3)=27 \text{ and } 9(-3)=-27$$

Also, any number times 0 always results in 0. If a problem consists of several multipliers, the result is negative if it contains an odd number of negative factors, and the result is positive if it contains an even number of negative factors. For example,

$$(-1)(-1)(-1)(-1)=1 \text{ and } (-1)(-1)(-1)(-1)(-1)=-1$$

These two problems are also examples of repeated multiplication, which can be written in a more compact notation using exponents. The first example can be written as $(-1)^4=1$, and the second example can be written as $(-1)^5=-1$.

Both are exponential expressions; −1 is the base in both instances, and 4 and 5 are the respective exponents. Note that a negative number raised to an odd power is always negative, and a negative number raised to an even power is always positive. Also, $(-1)^4$ is not the same as -1^4. In the first expression, the negative is included in the parentheses, but it is not in the second expression. The second expression is found by evaluating 1^4 first to get 1 and then by applying the negative sign to obtain −1.

Similar rules apply within division. First, consider some vocabulary. When dividing 14 by 2, it can be written in the following ways: $14\div 2=7$ or $\frac{14}{2}=7$. 14 is the **dividend,** 2 is the **divisor**, and 7 is the **quotient.** If two numbers in a division problem have the same sign, the quotient is positive. If two numbers in a division problem have different signs, the quotient is negative. For example,

$$14\div(-2)=-7 \text{ and } -14\div(-2)=7$$

To check division, multiply the quotient by the divisor to obtain the dividend. Also, remember that 0 divided by any number is equal to 0. However, any number divided by 0 is undefined. It just does not make sense to divide a number by 0 parts.

If more than one operation is to be completed in a problem, follow the Order of Operations. The mnemonic device, PEMDAS, states the order in which addition, subtraction, multiplication, and division need to be done. It also includes when to evaluate operations within grouping symbols and when to incorporate exponents. PEMDAS, which some remember by thinking "please excuse my dear Aunt Sally," refers to parentheses, exponents, multiplication, division, addition, and subtraction. First, complete any operation within parentheses or any other grouping symbol like brackets, braces, or absolute value symbols.

Note that this does not refer to when parentheses are used to represent multiplication like $(2)(5)$ because in such cases, an operation is not within parentheses like it is in $(2 \cdot 5)$. Then, any exponents must be computed. Next, multiplication and division are performed from left to right. Finally, addition and subtraction are performed from left to right.

The following is an example in which the operations within the parentheses need to be performed first, so the order of operations must be applied to the exponent, subtraction, addition, and multiplication within the grouping symbol:

$$9-3(3^2-3+4\cdot 3)$$

$$\begin{aligned} &9-3(3^2-3+4\cdot 3) \\ &= 9-3(9-3+12) \\ &= 9-3(18) \\ &= 9-54 \\ &= -45 \end{aligned}$$

Work within the parentheses first

Once the rules for integers are understood, move on to learning how to perform operations with fractions and decimals. Recall that a rational number can be written as a fraction and can be converted to a decimal through division. If a rational number is negative, the rules for adding, subtracting, multiplying, and dividing integers must be used. If a rational number is in fraction form, performing addition, subtraction, multiplication, and division is more complicated than when working with integers. First, consider addition. To add two fractions having the same denominator, add the numerators and then reduce the fraction. When an answer is a fraction, it should always be in lowest terms. **Lowest terms** means that every common factor, other than 1, between the numerator and denominator is divided out. For example:

$$\frac{2}{8}+\frac{4}{8}=\frac{6}{8}$$

$$\frac{6\div 2}{8\div 2}=\frac{3}{4}$$

Both the numerator and denominator of $\frac{6}{8}$ have a common factor of 2, so 2 is divided out of each number to put the fraction in lowest terms. If denominators are different in an addition problem, the fractions must be converted to have common denominators. The **least common denominator (LCD)** of all the given denominators must be found, and this value is equal to the **least common multiple (LCM)** of the denominators. This non-zero value is the smallest number that is a multiple of both denominators. Then, rewrite each original fraction as an equivalent fraction using the new denominator. Once in this form, apply the process of adding with like denominators. For example, consider $\frac{1}{3}+\frac{4}{9}$. The LCD is 9 because it is the smallest multiple of both 3 and 9. The fraction $\frac{1}{3}$ must be rewritten with 9 as its denominator. Therefore, multiply both the numerator and denominator by 3. Multiplying by $\frac{3}{3}$ is the same as multiplying by 1, which does not change the value of the fraction. Therefore, an equivalent fraction is $\frac{3}{9}$, and $\frac{1}{3}+\frac{4}{9}=\frac{3}{9}+\frac{4}{9}=\frac{7}{9}$, which is in lowest terms. Subtraction is performed in a similar manner; once the denominators are equal, the numerators are then subtracted.

The following is an example of addition of a positive and a negative fraction:

$$-\frac{5}{12}+\frac{5}{9}$$

$$-\frac{5\times 3}{12\times 3}+\frac{5\times 4}{9\times 4}$$

$$-\frac{15}{36}+\frac{20}{36}=\frac{5}{36}$$

Common denominators are not used in multiplication and division. To multiply two fractions, multiply the numerators together and the denominators together. Then, write the result in lowest terms. For example:

$$\frac{2}{3}\times\frac{9}{4}=\frac{18}{12}=\frac{3}{2}$$

Alternatively, the fractions could be factored first to cancel out any common factors before performing the multiplication. For example:

$$\frac{2}{3}\times\frac{9}{4}=\frac{2}{3}\times\frac{3\times 3}{2\times 2}=\frac{3}{2}$$

This second approach is helpful when working with larger numbers, as common factors might not be obvious. Multiplication and division of fractions are related because the division of two fractions is changed into a multiplication problem. This means that dividing a fraction by another fraction is the same as multiplying the first fraction by the reciprocal of the second fraction, so that second fraction must be inverted, or "flipped," to be in reciprocal form. For example:

$$\frac{11}{15}\div\frac{3}{5}=\frac{11}{15}\times\frac{5}{3}=\frac{55}{45}=\frac{11}{9}$$

The fraction $\frac{5}{3}$ is the reciprocal of $\frac{3}{5}$.

It is possible to multiply and divide numbers containing a mix of integers and fractions. In this case, convert the integer to a fraction by placing it over a denominator of 1. For example, a division problem involving an integer and a fraction is:

$$3\div\frac{1}{2}$$

$$\frac{3}{1}\times\frac{2}{1}$$

$$\frac{6}{1} = 6$$

Finally, when performing operations with rational numbers that are negative, the same rules apply as when performing operations with integers. For example, a negative fraction multiplied by a negative fraction results in a positive value, and a negative fraction subtracted from a negative fraction results in a negative value.

Operations can be performed on rational numbers in decimal form. Recall that to write a fraction as an equivalent decimal expression, divide the numerator by the denominator.

For example:

$$\frac{1}{8} = 1 \div 8 = 0.125$$

With the case of decimals, it is important to keep track of place value. To add decimals, make sure the decimal places are in alignment and add vertically. If the numbers do not line up because there are extra or missing place values in one of the numbers, then zeros may be used as placeholders. For example, $0.123 + 0.23$ becomes:

$$\begin{array}{r} 0.123 \\ +\underline{0.230} \\ 0.353 \end{array}$$

Subtraction is done the same way. Multiplication and division are more complicated. To multiply two decimals, place one on top of the other as in a regular multiplication process and do not worry about lining up the decimal points. Then, multiply as with whole numbers, ignoring the decimals. Finally, in the solution, insert the decimal point as many places to the left as there are total decimal values in the original problem. Here is an example of a decimal multiplication problem:

0.52	***2 decimal places***
x 0.2	***1 decimal place***
0.104	***3 decimal places***

The answer to 52 times 2 is 104, and because there are three decimal values in the problem, the decimal point is positioned three units to the left in the answer.

The decimal point plays an integral role throughout the whole problem when dividing with decimals. First, set up the problem in a long division format. If the divisor is not an integer, move the decimal to the right as many units as needed to make it an integer. The decimal in the dividend must be moved to the right the same number of places to maintain equality.

Then, complete division normally. Here is an example of long division with decimals:

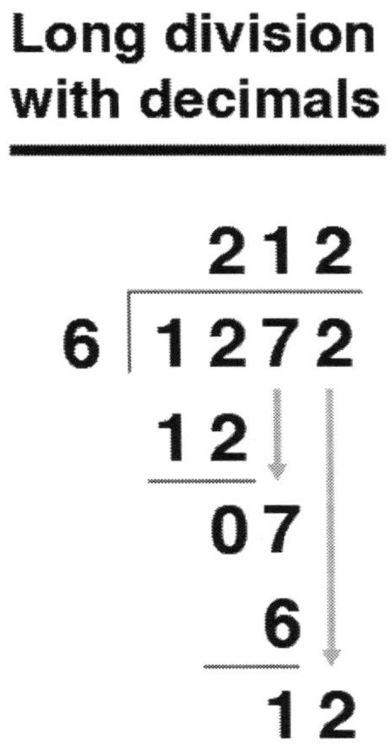

The decimal point in 0.06 needed to move two units to the right to turn it into an integer (6), so it also needed to move two units to the right in 12.72 to make it 1,272. The result is 212, and remember that a division problem can always be checked by multiplying the answer by the divisor to see if the result is equal to the dividend.

Sometimes it is helpful to round answers that are in decimal form. First, find the place to which the rounding needs to be done. Then, look at the digit to the right of it. If that digit is 4 or less, the number in the place value to its left stays the same, and everything to its right becomes a 0. This process is known as rounding down. If that digit is 5 or higher, the number in the place value to its left increases by 1, and every number to its right becomes a 0. This is called rounding up. Excess 0s at the end of a decimal can be dropped. For example, 0.145 rounded to the nearest hundredth place would be rounded up to 0.15, and 0.145 rounded to the nearest tenth place would be rounded down to 0.1.

Another operation that can be performed on rational numbers is the square root. Dealing with real numbers only, the **positive square root** of a number is equal to one of the two repeated positive factors of that number. For example:

$$\sqrt{49} = \sqrt{7 \cdot 7} = 7$$

A **perfect square** is a number that has a whole number as its square root. Examples of perfect squares are 1, 4, 9, 16, 25, etc. If a number is not a perfect square, an approximation can be used with a calculator. For example, $\sqrt{67} = 8.185$, rounded to the nearest thousandth place. Taking the square root of a fraction that includes perfect squares involves breaking up the problem into the square root of the numerator separate from the square root of the denominator. For example:

$$\sqrt{\frac{16}{25}} = \frac{\sqrt{16}}{\sqrt{25}} = \frac{4}{5}$$

If the fraction does not contain perfect squares, a calculator can be used. Therefore, $\sqrt{\frac{2}{5}} = 0.632$, rounded to the nearest thousandth place. A common application of square roots involves the Pythagorean theorem. Given a right triangle, the sum of the squares of the two legs equals the square of the hypotenuse.

For example, consider the following right triangle:

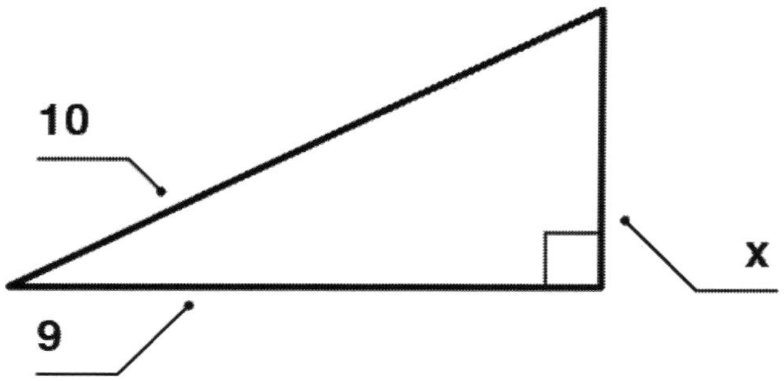

The missing side, x, can be found using the Pythagorean theorem.

$$9^2 + x^2 = 10^2$$

$$81 + x^2 = 100$$

$$x^2 = 19$$

To solve for x, take the square root of both sides. Therefore, $x = \sqrt{19} = 4.36$, which has been rounded to two decimal places.

In addition to the square root, the cube root is another operation. If a number is a **perfect cube**, the cube root of that number is equal to one of the three repeated factors. For example:

$$\sqrt[3]{27} = \sqrt[3]{3 \cdot 3 \cdot 3} = 3$$

A negative number has a cube root, which will also be a negative number. For example:

$$\sqrt[3]{-27}$$

$$\sqrt[3]{(-3)(-3)(-3)} = -3$$

Similar to square roots, if the number is not a perfect cube, a calculator can be used to find an approximation. Therefore, $\sqrt[3]{\frac{2}{3}} = 0.873$, rounded to the nearest thousandth place.

Higher-order roots also exist. The number relating to the root is known as the **index.** Given the following root, $\sqrt[3]{64}$, 3 is the index, and 64 is the **radicand.** The entire expression is known as the **radical.** Higher-order roots exist when the index is larger than 3. They can be broken up into two groups: even and odd roots. Even roots, when the index is an even number, follow the properties of square roots. They are found by finding the number that, when multiplied by itself the number of times indicated by the index, results in the radicand.

For example, the fifth root of 32 is equal to 2 because:

$$\sqrt[5]{32} = \sqrt[5]{2 \cdot 2 \cdot 2 \cdot 2 \cdot 2} = 2$$

Odd roots, when the index is an odd number, follow the properties of cube roots. A negative number has an odd root. Similarly, an odd root is found by finding the single factor that is repeated that many times to obtain the radicand. For example, the 4th root of 81 is equal to 3 because $3^4 = 81$. This radical is written as $\sqrt[4]{81} = 3$. Higher-order roots can also be evaluated on fractions and decimals, for example, because:

$$\left(\frac{2}{7}\right)^4 = \frac{16}{2{,}401}, \sqrt[4]{\frac{16}{2{,}401}} = \frac{2}{7}$$

and because:

$$(0.1)^5 = 0.00001, \sqrt[5]{0.00001} = 0.1$$

Sometimes, when performing operations in rational numbers, it might be helpful to round the numbers in the original problem to get a rough estimate of what the answer should be. For example, if you walked into a grocery store and had a $20 bill, you could round each item to the nearest dollar and add up all the items to make sure that you will have enough money when you check out. This process involves obtaining an estimation of what the exact total would be. In other situations, it might be helpful to round to the nearest $10 amount or $100 amount.

Front-end rounding might be helpful as well in many situations. In this type of rounding, the first digit of a number is rounded to the highest possible place value. Then, all digits following the first become 0. Consider a situation in which you are at the furniture store and want to estimate your total on three pieces of furniture that cost $434.99, $678.99, and $129.99. Front-end rounding would round these three amounts to $500, $700, and $200. Therefore, the estimate of your total would be $\$500 + \$700 + \$200 = \$1{,}400$, compared to the exact total of $1,243.97. In this situation, the estimate is not that far off the exact answer.

Rounding is useful both for approximation when an exact answer is not needed and for comparison when an exact answer is needed. For instance, if you had a complicated set of operations to complete and your estimate was $1,000, if you obtained a calculated answer of $100,000, something is off. You might want to check your work to see if you made a mistake because an estimate should not be that different from an exact answer. Estimates can also be helpful with square roots. If the square root of a number is unknown, then you can use the closest perfect square to help you approximate. For example, $\sqrt{50}$ is not equal to a whole number, but 50 is close to 49, which is a perfect square, and $\sqrt{49} = 7$. Therefore, $\sqrt{50}$ is a little bit larger than 7. The actual approximation, rounded to the nearest thousandth, is 7.071.

Ordering and Comparing Rational Numbers

Ordering rational numbers is a way to compare two or more different numerical values. Determining whether two amounts are equal, or if one is less than or greater than the other is the basis for comparing both positive and negative numbers. Also, a group of numbers can be compared by ordering them from the smallest value to the largest value. A few symbols are necessary to use when ordering rational numbers. The equals sign, $=$, shows that the two quantities on either side of the symbol have the same value. For example, $\frac{12}{3} = 4$ because both values are equivalent.

Another symbol that is used to compare numbers is $<$, which represents "less than." With this symbol, the smaller number is placed on the left and the larger number is placed on the right. Always remember that the symbol's "mouth" opens up to the larger number. When comparing negative and positive numbers, it is important to remember that the number occurring to the left on the number line is always smaller and is placed to the left of the symbol. This idea might seem confusing because some values could appear at first glance to be larger, even though they are not. For example, $-5 < 4$ is read "negative 5 is less than 4." Here is an image of a number line to help visualize relationships:

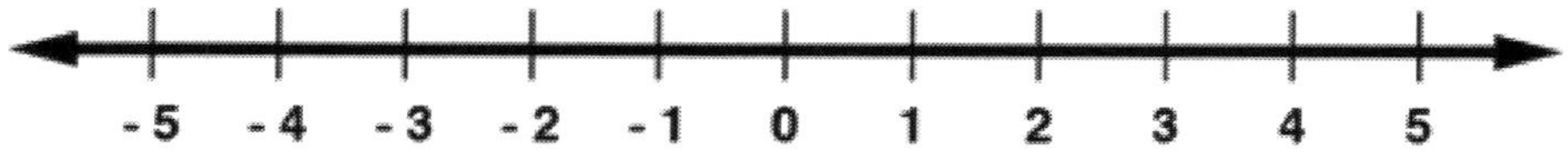

The symbol $\leq$ represents "less than or equal to," and it joins $<$ with equality. Therefore, both $-5 \leq 4$ and $-5 \leq -5$ are true statements and "-5 is less than or equal to both 4 and -5." Other symbols are $>$ and $\geq$, which represent "greater than" and "greater than or equal to." Both $4 \geq -1$ and $-1 \geq -1$ are correct ways to use these symbols.

Here is a chart of these four inequality symbols:

Symbol	Definition
$<$	less than
$\leq$	less than or equal to
$>$	greater than
$\geq$	greater than or equal to

Comparing integers is a straightforward process, especially when using the number line, but the comparison of decimals and fractions is not as obvious. When comparing two non-negative decimals, compare digit by digit, starting from the left. The larger value contains the first larger digit. For example, 0.1456 is larger than 0.1234 because the value 4 in the hundredths place in the first decimal is larger than the value 2 in the hundredths place in the second decimal. When comparing a fraction with a decimal, convert the fraction to a decimal and then compare in the same manner. Finally, there are a few options when comparing multiple fractions. If two non-negative fractions have the same denominator, the fraction with the larger numerator is the larger value. If they have the same numerator but different denominators (such as $\frac{3}{5}$ and $\frac{3}{10}$), the fraction whose denominator is a smaller number (in this case, the $\frac{3}{5}$ because 5 is less than 10) is larger.

If the two fractions have different numerators and denominators, they can be converted to equivalent fractions with a common denominator to be compared, or they can be converted to decimals to be compared. When comparing two negative decimals or fractions, a different approach must be used. It is important to remember that the smaller number exists to the left on the number line. Therefore, when comparing two negative decimals by place value, the number with the larger first place value is smaller due to the negative sign. Whichever value is closer to 0 is larger. For instance, -0.456 is larger than -0.498 because of the values in the hundredth places. If two negative fractions have the same denominator, the fraction with the larger numerator is smaller because of the negative sign.

Solving Real-World One- or Multi-Step Problems with Rational Numbers

One-step problems take only one mathematical step to solve. For example, solving the equation $5x = 45$ is a one-step problem because the one step of dividing both sides of the equation by 5 is the only step necessary to obtain the solution $x = 9$. The **multiplication principle of equality** is the one step used to isolate the variable. The equation is of the form $ax = b$, where a and b are rational numbers. Similarly, the **addition principle of equality** could be the one step needed to solve a problem. In this case, the equation would be of the form $x + a = b$ or $x - a = b$, for real numbers a and b.

A multi-step problem requires more than one step to find the solution, or it might consist of solving more than one equation. An equation that involves both the addition principle and the multiplication principle is a two-step problem, and an example of such an equation is:

$$2x - 4 = 5$$

To solve, add 4 to both sides and then divide both sides by 2. An example of a two-step problem involving two separate equations is:

$$y = 3x, \qquad 2x + y = 4$$

The two equations form a system that must be solved together in two variables. The system can be solved by the substitution method. Since y is already solved for in terms of x, replace y with $3x$ in the equation $2x + y = 4$, resulting in:

$$2x + 3x = 4$$

Therefore, $5x = 4$ and $x = \frac{4}{5}$. Because there are two variables, the solution requires a value for both x and for y. Substitute $x = \frac{4}{5}$ into either original equation to find y. The easiest choice is $y = 3x$. Therefore:

$$y = 3 \times \frac{4}{5} = \frac{12}{5}$$

The solution can be written as the ordered pair $\left(\frac{4}{5}, \frac{12}{5}\right)$.

Real-world problems can be translated into both one-step and multi-step problems. In either case, the word problem must be translated from the verbal form into mathematical expressions and equations that can be solved using algebra. An example of a one-step real-world problem is the following: A cat weighs half as much as a dog living in the same house. If the dog weighs 14.5 pounds, how much does the cat weigh? To solve this problem, an equation can be used.

In any word problem, the first step is to define variables that represent the unknown quantities. For this problem, let x be equal to the unknown weight of the cat. Because two times the weight of the cat equals 14.5 pounds, the equation to be solved is: $2x = 14.5$. Use the multiplication principle to divide both sides by 2. Therefore, $x = 7.25$, and the cat weighs 7.25 pounds.

Most of the time, real-world problems require multiple steps. The following is an example of a multi-step problem: The sum of two consecutive page numbers is equal to 437. What are those page numbers? First, define the unknown quantities. If x is equal to the first page number, then $x + 1$ is equal to the next page number because they are consecutive integers. Their sum is equal to 437. Putting this information together results in the equation:

$$x + x + 1 = 437$$

To solve, first collect like terms to obtain:

$$2x + 1 = 437$$

Then, subtract 1 from both sides and then divide by 2. The solution to the equation is $x = 218$. Therefore, the two consecutive page numbers that satisfy the problem are 218 and 219.

It is always important to make sure that answers to real-world problems make sense. For instance, it should be a red flag if the solution to this same problem resulted in decimals, which would indicate the need to check the work. Page numbers are whole numbers; therefore, if decimals are found to be answers, the solution process should be double-checked for mistakes.

Factorization

Factorization is the process of breaking up a mathematical quantity, such as a number or polynomial, into a product of two or more factors. For example, a factorization of the number 16 is:

$$16 = 8 \times 2$$

If multiplied out, the factorization results in the original number. A **prime factorization** is a specific factorization when the number is factored completely using prime numbers only. For example, the prime factorization of 16 is:

$$16 = 2 \times 2 \times 2 \times 2$$

A factor tree can be used to find the prime factorization of any number.

Within a factor tree, pairs of factors are found until no other factors can be used, as in the following factor tree of number 84:

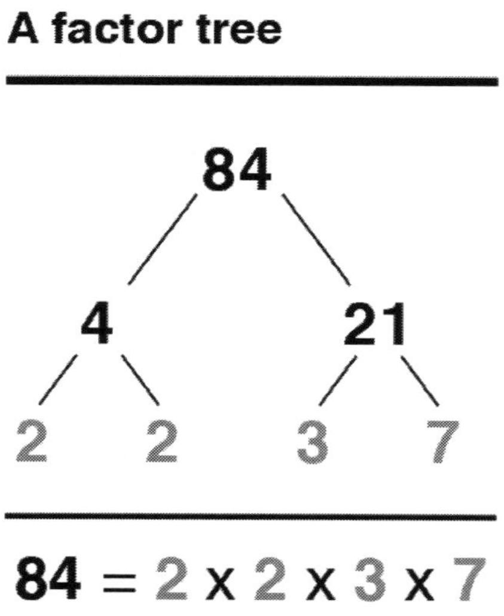

It first breaks 84 into 21×4, which is not a prime factorization. Then, both 21 and 4 are factored into their primes. The final numbers on each branch consist of the numbers within the prime factorization. Therefore:

$$84 = 2 \times 2 \times 3 \times 7$$

Factorization can be helpful in finding greatest common divisors and least common denominators.

Also, a factorization of an algebraic expression can be found. Throughout the process, a more complicated expression can be decomposed into products of simpler expressions. To factor a polynomial, first determine if there is a greatest common factor. If there is, factor it out. For example, $2x^2 + 8x$ has a greatest common factor of $2x$ and can be written as $2x(x + 4)$. Once the greatest common monomial factor is factored out, if applicable, count the number of terms in the polynomial. If there are two terms, is it a difference of squares, a sum of cubes, or a difference of cubes?

If so, the following rules can be used:

$$a^2 - b^2 = (a + b)(a - b)$$

$$a^3 + b^3 = (a + b)(a^2 - ab + b^2)$$

$$a^3 - b^3 = (a - b)(a^2 + ab + b^2)$$

If there are three terms, and if the trinomial is a perfect square trinomial, it can be factored into the following:

$$a^2 + 2ab + b^2 = (a + b)^2$$

$$a^2 - 2ab + b^2 = (a - b)^2$$

If not, try factoring into a product of two binomials in the form of:

$$(x + p)(x + q)$$

For example, to factor $x^2 + 6x + 8$, determine what two numbers have a product of 8 and a sum of 6. Those numbers are 4 and 2, so the trinomial factors into:

$$(x + 2)(x + 4)$$

Finally, if there are four terms, try factoring by grouping. First, group terms together that have a common monomial factor. Then, factor out the common monomial factor from the first two terms. Next, look to see if a common factor can be factored out of the second set of two terms that results in a common binomial factor. Finally, factor out the common binomial factor of each expression, for example:

$$xy - x + 5y - 5$$

$$x(y - 1) + 5(y - 1)$$

$$(y - 1)(x + 5)$$

After the expression is completely factored, check the factorization by multiplying it out; if this results in the original expression, then the factoring is correct. Factorizations are helpful in solving equations that consist of a polynomial set equal to zero. If the product of two algebraic expressions equals zero, then at least one of the factors is equal to zero. Therefore, factor the polynomial within the equation, set each factor equal to zero, and solve. For example, $x^2 + 7x - 18 = 0$ can be solved by factoring into:

$$(x + 9)(x - 2) = 0$$

Set each factor equal to zero, and solve to obtain $x = -9$ and $x = 2$.

Identifying Integers

Integers include zero, and both positive and negative numbers with no fractional component. Examples of integers are −3, 5, 120, −47, and 0. Numbers that are not integers include 1.3333, ½, −5.7, and 4 ½. Integers can be used to describe different real-world situations. If a scuba diver were to dive 50 feet down into the ocean, his position can be described as −50, in relation to sea level. If, while traveling in Denver, Colorado, a car has an elevation reading of 2300 feet, the integer 2300 can be used to describe the feet above sea level. Integers can be used in many different ways to describe situations with whole numbers and zero.

Integers can also be added and subtracted as situations change. If the temperature in the morning is 45 degrees, and it dropped to 33 degrees in the afternoon, the difference can be found by subtracting the integers 45 and 33 to get a change of 12 degrees. If a submarine was at a depth of 100 feet below sea level, then rose 35 feet, the new depth can be found by adding −100 to 35. The following equation can be used to model the situation with integers:

$$-100 + 35 \quad -65$$

The answer of −65 reveals the new depth of the submarine, 65 feet below sea level.

Recognizing Rational Exponents

A **rational number** is any number that can be written as a fraction of two integers. Examples of rational numbers include $\frac{1}{2}, \frac{5}{4}$, and 8. The number 8 is rational because it can be expressed as a fraction: $\frac{8}{1} = 8$.

Rational exponents are used to express the root of a number raised to a specific power. For example, $3^{\frac{1}{2}}$ has a base of 3 and rational exponent of $\frac{1}{2}$. The square root of 3 raised to the first power can be written as $\sqrt[2]{3^1}$. Any number with a rational exponent can be written this way. The **numerator**, or number on top of the fraction, becomes the whole number exponent and the **denominator**, or bottom number of the fraction, becomes the root. Another example is $4^{\frac{3}{2}}$. It can be rewritten as the square root of four to the third power, or $\sqrt[2]{4^3}$. To simplify this, first solve for 4 to the third power:

$$4^3 = 4 \times 4 \times 4 = 64$$

Then take the square root of 64, written as $\sqrt[2]{64}$, which yields an answer of 8. Another way of stating the answer would be 4 to power of $\frac{3}{2}$ is eight, or that 4 to the power of $\frac{3}{2}$ is the square root of 4 cubed,

$$\sqrt[2]{4}^{\,3} = 2^3 = 2 \times 2 \times 2 = 8$$

Understanding Vectors

A **vector** is something that has both magnitude and direction. A vector may sometimes be represented by a ray that has a length, for its magnitude, and a direction. As the magnitude of the vector increases, the length of the ray changes. The direction of the ray refers to the way that the magnitude is applied. The following image shows the placement and parts of a vector.

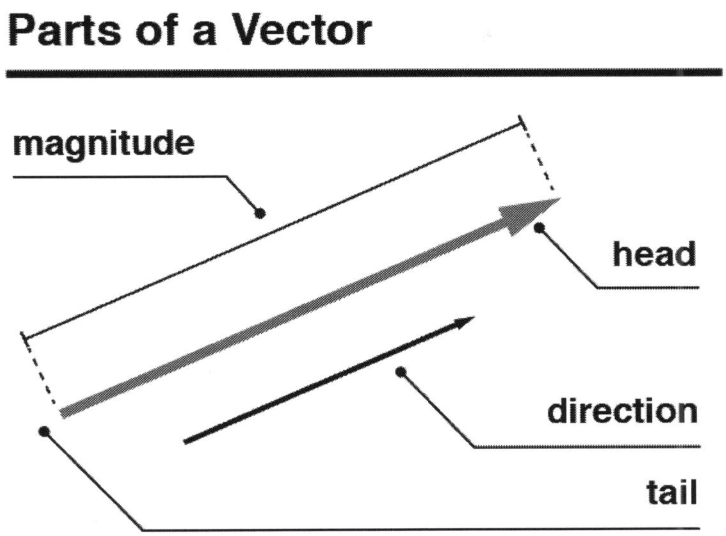

Examples of vector quantities include force and velocity. **Force** is a vector quantity because applying force requires magnitude, which is the amount of force, and a direction that the force is applied. **Velocity** is a vector because it has a magnitude, or speed that an object travels, and also the direction that the object is traveling in. Vectors can be added together by placing the tail of the second at the head of the first. The resulting vector is found by starting at the first tail and ending at the second head. An example of this is show in the following picture.

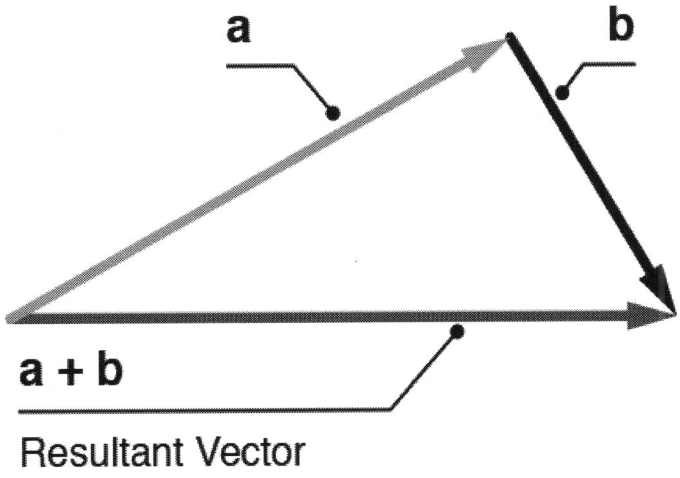

Subtraction can also be done with vectors by adding the inverse of the second vector. The inverse is found by reversing the direction of the vector. Then addition can take place just as described above, but using the inverse instead of the original vector. Scalar multiplication can also be done with vectors. This multiplication changes the magnitude of the vector by the **scalar,** or number. For example, if the length is described as 4, then scalar multiplication is used to multiply by 2, so the vector magnitude becomes 8. The direction of the vector is left unchanged because scalar does not include direction.

Vectors may also be described using coordinates on a plane, such as (5, 2). This vector would start at a point and move to the right 5 and up 2. The two coordinates describe the horizontal and vertical components of the vector. The starting point in relation to the coordinates is the tail, and the ending point is the head.

Creating Matrices

A **matrix** is an arrangement of numbers in rows and columns. Matrices are used to work with vectors and transform them. One example is a system of linear equations. Matrices can represent a system and be used to transform and solve the system. An important connection between scalars, vectors, and matrices is this: scalars are only numbers, vectors are numbers with magnitude and direction, and matrices are an array of numbers in rows and columns. The rows run from left to right and the columns run from top to bottom. When describing the dimensions of a matrix, the number of rows is stated first, and the number of columns is stated second. The following matrix has two rows and three columns, referred to as a 2×3 matrix:

$$\begin{bmatrix} 3 & 5 & 7 \\ 4 & 2 & 8 \end{bmatrix}$$

A number in a matrix can be found by describing its location. For example, the number in row two, column three is 8. In row one, column two, the number 5 is found.

Operations can be performed on matrices, just as they can on vectors. Scalar multiplication can be performed on matrices and it will change the magnitude, just as with a vector. A scalar multiplication problem using a 2×2 matrix looks like the following:

$$3 \times \begin{bmatrix} 4 & 5 \\ 8 & 3 \end{bmatrix}$$

The scalar of 3 is multiplied by each number to form the resulting matrix:

$$\begin{bmatrix} 12 & 15 \\ 24 & 9 \end{bmatrix}$$

Matrices can also be added and subtracted. For these operations to be performed, the matrices must be the same dimensions. Other operations that can be performed to manipulate matrices are multiplication, division, and transposition. **Transposing** a matrix means to switch the rows and columns. If the original matrix has two rows and three columns, then the transposed matrix has three rows and two columns.

Algebra

Solving, Graphing, and Modeling Multiple Types of Expressions

When presented with a real-world problem, the first step is to determine what unknown quantity must be solved for. Use a **variable**, such as x or t, to represent that unknown quantity. Sometimes there can be two or more unknown quantities. In this case, either choose an additional variable, or if a relationship exists between the unknown quantities, express the other quantities in terms of the original variable. After choosing the variables, form algebraic expressions and/or equations that represent the verbal statement in the problem. The following table shows examples of vocabulary used to represent the different operations:

Addition	Sum, plus, total, increase, more than, combined, in all
Subtraction	Difference, less than, subtract, reduce, decrease, fewer, remain
Multiplication	Product, multiply, times, part of, twice, triple
Division	Quotient, divide, split, each, equal parts, per, average, shared

The combination of operations and variables form both mathematical expression and equations. The differences between expressions and equations are that there is no equals sign in an expression, and that expressions are evaluated to find an unknown quantity, while equations are solved to find an unknown quantity. Also, inequalities can exist within verbal mathematical statements. Instead of a statement of equality, expressions state quantities are *less than*, *less than or equal to*, *greater than*, or *greater than or equal to*. Another type of inequality is when a quantity is said to be not equal to another quantity ($\neq$).

The steps for solving inequalities in one variable are the same steps for solving equations in one variable. The addition and multiplication principles are used. However, to maintain a true statement when using the $<, \leq, >$, and $\geq$ symbols, if a negative number is either multiplied times both sides of an inequality or divided from both sides of an inequality, the sign must be flipped. For instance, consider the following inequality:

$$3 - 5x \leq 8$$

First, 3 is subtracted from each side to obtain $-5x \leq 5$. Then, both sides are divided by -5, while flipping the sign, to obtain $x \geq -1$. Therefore, any real number greater than or equal to -1 satisfies the original inequality.

Adding and Subtracting Linear Algebraic Expressions

To add and subtract linear algebraic expressions, you must combine like terms. Like terms are terms that have the same variable with the same exponent. In the following example, the x-terms can be added because the variable and exponent are the same. These terms add to be $9x$. Terms without a variable component are called constants. These terms will add to be nine.

Example: Add $(3x - 5) + (6x + 14)$

$3x - 5 + 6x + 14$ Rewrite without parentheses

$3x + 6x - 5 + 14$ Commutative property of addition

$9x + 9$ Combine like terms

When subtracting linear expressions, be careful to add the opposite when combining like terms. Do this by distributing -1, which is multiplying each term inside the second parenthesis by negative one. Remember that distributing -1 changes the sign of each term.

Example: Subtract $(17x + 3) - (27x - 8)$

$17x + 3 - 27x + 8$ Distributive Property

$17x - 27x + 3 + 8$ Commutative property of addition

$-10x + 11$ Combine like terms

Example: Simplify by adding or subtracting:

$(6m + 28z - 9) + (14m + 13) - (-4z + 8m + 12)$

$6m + 28z - 9 + 14m + 13 + 4z - 8m - 12$	Distributive Property
$6m + 14m - 8m + 28z + 4z - 9 + 13 - 12$	Commutative Property of Addition
$12m + 32z - 8$	Combine like terms

Solving Problems Using Numerical and Algebraic Expressions

Translating sentences describing relationships between variables and constants to algebraic expressions and equations involves recognizing key words that represent mathematical operations. This process is known as **modeling.** For simplicity, let x be the variable, or the unknown quantity. Statements that include the four operations addition, subtraction, multiplication, and division exist in sentences that model linear relationships. For example, words and phrases that represent addition are "sum," "more than," and "increased by." Words and phrases that represent subtraction are "minus," "decreased by," "subtracted from," "difference," "less", "fewer than," and "less than." Words and phrases that represent multiplication are "times," "product of," "twice," "double," and "triple." Finally, words and phrases that represent division are "divided by," "quotient," and "reciprocal."

Most of the time, these words and phrases are combined to represent expressions that deal with one or more operation. For example, "ten subtracted from nine times a number" would be represented as $9x - 10$, and "the quotient of a number and 7 increased by 8" would be represented as $\frac{x}{7} + 8$. The word problems that typically use these expressions will have a statement of equality. For instance, the problem could say "ten subtracted from nine times a number equals 20; find that number." In this case, the algebraic expression shown previously would be set equal to 20 and then solved for x.

$$9x - 10 = 20$$

Add ten to both sides and then divide by 9 to get the solution $x = \frac{30}{9}$, which reduces to $x = \frac{10}{3}$.

Other types of expressions, besides linear expressions, can be the results of modeling. If the variable is raised to a power other than 1, the result is a **polynomial expression**. The path of an object thrown up into the air is a common example of this. The graph of an object represents an upside-down parabola, which is modeled by an equation of the type $y = -ax^2$. In this case, *a* represents the height of the object at its highest point before coming back down to the ground, and the negative sign shows that the parabola is upside down.

Here is the graph of a parabola:

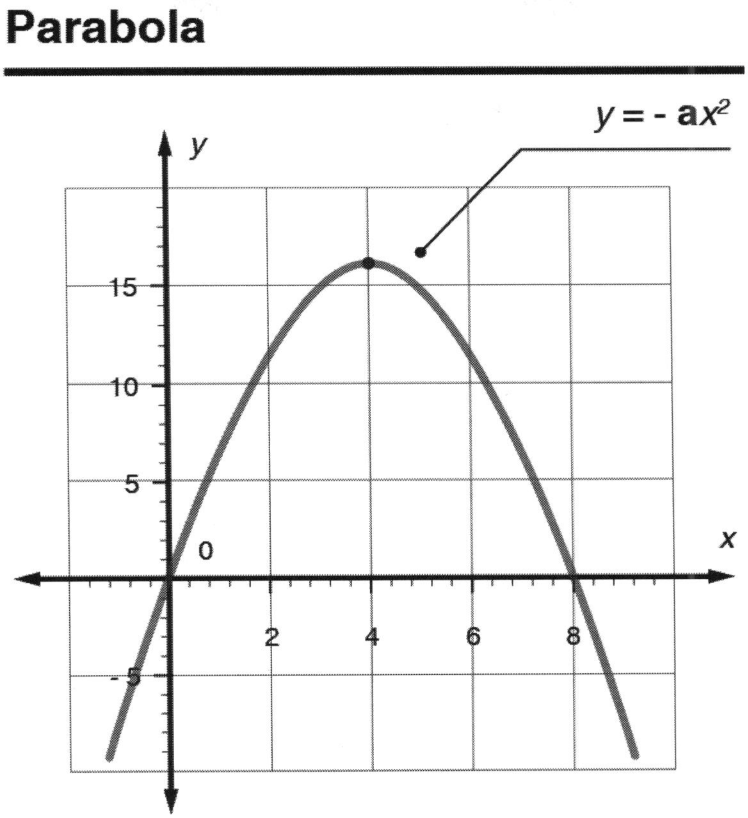

Evaluating and Simplifying Algebraic Expressions

Given an algebraic expression, students may be asked to evaluate for given values of variable(s). In doing so, students will arrive at a numerical value as an answer. For example:

$$\text{Evaluate } a - 2b + ab \; for \; a = 3 \text{ and } b = -1$$

To evaluate an expression, the given values should be substituted for the variables and simplified using the order of operations. In this case:

$$(3) - 2(-1) + (3)(-1)$$

Parentheses are used when substituting.

Given an algebraic expression, students may be asked to simplify the expression. For example, simplify:

$$5x^2 - 10x + 2 - 8x^2 + x - 1$$

Simplifying algebraic expressions requires combining like terms. A term is a number, variable, or product of a number and variables separated by addition and subtraction. The terms in the above expression are: $5x^2, -10x, 2, -8x^2, x,$ and -1. Like terms have the same variables raised to the same powers (exponents). To combine like terms, the coefficients (numerical factor of the term including sign) are added, while the variables and their powers are kept the same. The example above simplifies to:

$$-3x^2 - 9x + 1$$

Generating Equivalent Expressions

Two algebraic expressions are equivalent if they represent the same value, even if they look different. To obtain an equivalent form of an algebraic expression, follow the laws of algebra. For instance, addition and multiplication are both commutative and associative. Therefore, terms in an algebraic expression can be added in any order and multiplied in any order. For instance, $4x + 2y$ is equivalent to $2y + 4x$ and:

$$y \times 2 + x \times 4$$

Also, the distributive law allows a number to be distributed throughout parentheses, as in the following:

$$a(b + c) = ab + ac$$

The expressions on both sides of the equals sign are equivalent. Collecting like terms is also important when working with equivalent forms because the simplest version of an expression is always the easiest one to work with.

An expression is not an equation; therefore, expressions cannot undergo multiplication, division, addition, or subtraction and still have equivalent expressions. These processes can only happen in equations when the same step is performed on both sides of the equals sign.

The **distributive property** ($a(b + c) = ab + ac$) is a way of taking a factor and multiplying it through a given expression in parentheses. Each term inside the parentheses is multiplied by the outside factor, eliminating the parentheses. The following example shows how to distribute the number 3 to all the terms inside the parentheses.

Example: Use the distributive property to write an equivalent algebraic expression:

$3(2x + 7y + 6)$	
$3(2x) + 3(7y) + 3(6)$	Distributive property
$6x + 21y + 18$	Simplify

Because $a - b$ can be written $a + (-b)$, the distributive property can be applied in the example below:

Example: Use the distributive property to write an equivalent algebraic expression.

$7(5m - 8)$	
$7[5m + (-8)]$	Rewrite subtraction as addition of -8
$7(5m) + 7(-8)$	Distributive property
$35m - 56$	Simplify

In the following example, note that the factor of 2 is written to the right of the parentheses but is still distributed as before.

Example: Use the distributive property to write an equivalent algebraic expression:

$(3m + 4x - 10)2$	
$(3m)2 + (4x)2 + (-10)2$	Distributive property
$6m + 8x - 20$	Simplify

Example: $-(-2m + 6x)$

In this example, the negative sign in front of the parentheses can be interpreted as $-1(-2m + 6x)$

$-1(-2m + 6x)$	
$-1(-2m) + (-1)(6x)$	Distributive property
$2m - 6x$	Simplify

Linear Equations

An **equation in one variable** is a mathematical statement where two algebraic expressions with one variable, usually x, are set equal. To solve the equation, the variable must be isolated on one side of the equals sign. The addition and multiplication principles of equality are used to isolate the variable. The **addition principle of equality** states that the same number can be added to or subtracted from both sides of an equation. Because the same value is being used on both sides of the equals sign, equality is maintained. For example, the equation $2x = 5x$ is equivalent to both $2x + 3 = 5x + 3$, and:

$$2x - 5 = 5x - 5$$

This principle can be used to solve the following equation:

$$x + 5 = 4$$

The variable x must be isolated, so to move the 5 from the left side, subtract 5 from both sides of the equals sign. Therefore:

$$x + 5 - 5 = 4 - 5$$

So, the solution is $x = -1$. This process illustrates the idea of an **additive inverse** because subtracting 5 is the same as adding -5. Basically, add the opposite of the number that must be removed to both sides of the equals sign. The multiplication principle of equality states that equality is maintained when both sides of an equation are multiplied or divided by the same number.

For example, $4x = 5$ is equivalent to both $16x = 20$ and $x = \frac{5}{4}$. Multiplying both sides by 4 and dividing both sides by 4 maintains equality. Solving the equation $6x - 18 = 5$ requires the use of both principles. First, apply the addition principle to add 18 to both sides of the equals sign, which results in $6x = 23$. Then use the multiplication principle to divide both sides by 6, giving the solution $x = \frac{23}{6}$. Using the multiplication principle in the solving process is the same as involving a multiplicative inverse. A **multiplicative inverse** is a value that, when multiplied by a given number, results in 1. Dividing by 6 is the same as multiplying by $\frac{1}{6}$, which is both the reciprocal and multiplicative inverse of 6.

When solving linear equations, check the answer by plugging the solution back into the original equation. If the result is a false statement, something was done incorrectly during the solution procedure. Checking the example above gives the following:

$$6 \times \frac{23}{6} - 18$$

$$23 - 18 = 5$$

Therefore, the solution is correct.

Some equations in one variable involve fractions or the use of the distributive property. In either case, the goal is to obtain only one variable term and then use the addition and multiplication principles to isolate that variable. Consider the equation:

$$\frac{2}{3}x = 6$$

To solve for x, multiply each side of the equation by the reciprocal of $\frac{2}{3}$, which is $\frac{3}{2}$. This step results in $\frac{3}{2} \times \frac{2}{3}x = \frac{3}{2} \times 6$, which simplifies into the solution $x = 9$. Now consider the equation:

$$3(x + 2) - 5x = 4x + 1$$

Use the distributive property to clear the parentheses. Therefore, multiply each term inside the parentheses by 3. This step results in:

$$3x + 6 - 5x = 4x + 1$$

Next, collect like terms on the left-hand side. **Like terms** are terms with the same variable or variables raised to the same exponent(s). Only like terms can be combined through addition or subtraction.

After collecting like terms, the equation is:

$$-2x + 6 = 4x + 1$$

Finally, apply the addition and multiplication principles. Add $2x$ to both sides to obtain:

$$6 = 6x + 1$$

Then, subtract 1 from both sides to obtain $5 = 6x$. Finally, divide both sides by 6 to obtain the solution"

$$\frac{5}{6} = x$$

Two other types of solutions can be obtained when solving an equation in one variable. There could be no solution, or the solution set could contain all real numbers. Consider the equation:

$$4x = 6x + 5 - 2x$$

First, the like terms can be combined on the right to obtain:

$$4x = 4x + 5$$

Next, subtract $4x$ from both sides. This step results in the false statement $0 = 5$. There is no value that can be plugged into x that will ever make this equation true. Therefore, there is no solution. The solution procedure contained correct steps, but the result of a false statement means that no value satisfies the equation. The symbolic way to denote that no solution exists is $\emptyset$. Next, consider the equation:

$$5x + 4 + 2x = 9 + 7x - 5$$

Combining the like terms on both sides results in:

$$7x + 4 = 7x + 4$$

The left-hand side is exactly the same as the right-hand side. Using the addition principle to move terms, the result is $0 = 0$, which is always true. Therefore, the original equation is true for any number, and the solution set is all real numbers. The symbolic way to denote such a solution set is $\mathbb{R}$, or in interval notation, $(-\infty, \infty)$.

Algebraically Solving Linear Equations or Inequalities in One Variable
A *linear equation in one variable* can be solved using the following steps:

1. Simplify both sides of the equation by removing all parentheses, using the distributive property, and collecting all like terms.

2. Collect all variable terms on one side of the equation and all constant terms on the other side by adding the same quantity to or subtracting the same quantity from both sides.

3. Isolate the variable by either multiplying or dividing both sides of the equation by the same number.

4. Check the answer.

The only difference between solving linear inequalities versus equations is that when multiplying by a negative number or dividing by a negative number, the direction of the inequality symbol must be reversed.

If an equation contains multiple fractions, it might make sense to clear the equation of fractions first by multiplying all terms by the least common denominator. Also, if an equation contains several decimals, it might make sense to clear the decimals as well by multiplying times a factor of 10. If the equation has decimals in the hundredths place, multiply every term in the equation by 100.

Polynomial Equations

A polynomial is a mathematical expression containing addition, subtraction, or multiplication of one or more constants multiplied by variables raised to positive powers. A polynomial is considered expanded when there are no variables contained within parentheses, the distributive property has been carried out for any terms that were within parentheses, and like terms have been collected.

When working with polynomials, **like terms** are terms that contain the same variables with the same powers. For example, x^4y^5 and $9x^4y^5$ are like terms. The coefficients are different, but the same variables are raised to the same powers. When adding polynomials, only terms that are considered like terms can be added. When adding two like terms, just add the coefficients and leave the variables alone. This process uses the distributive property. For example:

$$x^4y^5 + 9x^4y^5$$

$$(1 + 9)x^4y^5$$

$$10x^4y^5$$

Therefore, when adding two polynomials, simply add the like terms together. Unlike terms cannot be combined.

Subtracting polynomials involves adding the opposite of the polynomial being subtracted. Basically, the sign of each term in the polynomial being subtracted is changed, and then the like terms are combined because it is now an addition problem. For example, consider the following:

$$6x^2 - 4x + 2 - (4x^2 - 8x + 1)$$

Add the opposite of the second polynomial to obtain:

$$6x^2 - 4x + 2 + (-4x^2 + 8x - 1)$$

Then, collect like terms to obtain:

$$2x^2 + 4x + 1$$

Multiplying polynomials involves using the product rule for exponents that:

$$b^m b^n = b^{m+n}$$

Basically, when multiplying expressions with the same base, just add the exponents. Multiplying a monomial by a monomial involves multiplying the coefficients together and then multiplying the variables together using the product rule for exponents. For instance:

$$8x^2y \times 4x^4y^2 = 32x^6y^3$$

When multiplying a monomial by a polynomial that is not a monomial, use the distributive property to multiply each term of the polynomial times the monomial. For example:

$$3x(x^2 + 3x - 4) = 3x^3 + 9x^2 - 12x$$

Finally, multiplying two polynomials when neither one is a monomial involves multiplying each term of the first polynomial by each term of the second polynomial. There are some shortcuts, given certain scenarios.

For instance, a binomial times a binomial can be found by using the **FOIL (Firsts, Outers, Inners, Lasts)** method shown here.

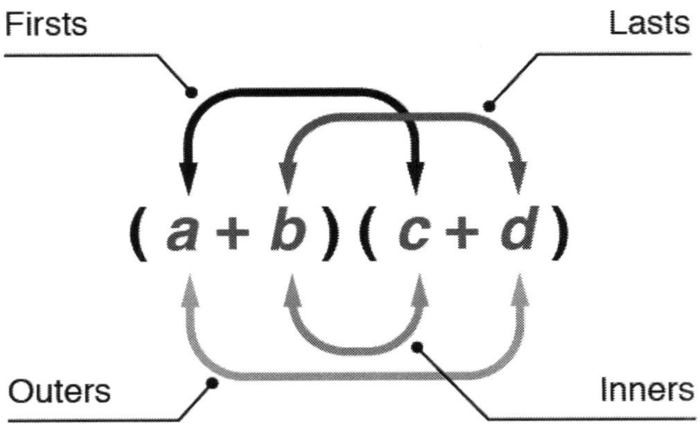

Finding the product of a sum and difference of the same two terms is simple because if it was to be foiled out, the outer and inner terms would cancel out. For instance:

$$(x + y)(x - y) = x^2 + xy - xy - y^2$$

Finally, the square of a binomial can be found using the following formula:

$$(a \pm b)^2 = a^2 \pm 2ab + b^2$$

Radical and Exponential Relationships

The *n*th root of a is given as $\sqrt[n]{a}$, which is called a **radical.** Typical values for *n* are 2 and 3, which represent the square and cube roots. In this form, *n* represents an integer greater than or equal to 2, and *a* is a real number. If *n* is even, *a* must be nonnegative, and if *n* is odd, *a* can be any real number. This radical can be written in exponential form as $a^{\frac{1}{n}}$. Therefore, $\sqrt[4]{15}$ is the same as $15^{\frac{1}{4}}$ and $\sqrt[3]{-5}$ is the same as $(-5)^{\frac{1}{3}}$.

In a similar fashion, the *n*th root of *a* can be raised to a power *m*, which is written as:

$$\left(\sqrt[n]{a}\right)^m$$

This expression is the same as:

$$\sqrt[n]{a^m}$$

For example:

$$\sqrt[2]{4^3} = \sqrt[2]{64} = 8 = \left(\sqrt[2]{4}\right)^3 = 2^3$$

Because $\sqrt[n]{a} = a^{\frac{1}{n}}$, both sides can be raised to an exponent of *m*, resulting in:

$$\left(\sqrt[n]{a}\right)^m = \sqrt[n]{a^m} = a^{\frac{m}{n}}$$

This rule allows:

$$\sqrt[2]{4^3} = \left(\sqrt[2]{4}\right)^3$$

$$4^{\frac{3}{2}} = (2^2)^{\frac{3}{2}}$$

$$2^{\frac{6}{2}} = 2^3 = 8$$

Negative exponents can also be incorporated into these rules. Any time an exponent is negative, the base expression must be flipped to the other side of the fraction bar and rewritten with a positive exponent. For instance:

$$2^{-3} = \frac{1}{2^3} = \frac{1}{8}$$

Therefore, two more relationships between radical and exponential expressions are:

$$a^{-\frac{1}{n}} = \frac{1}{\sqrt[n]{a}}$$

and

$$a^{-\frac{m}{n}} = \frac{1}{\sqrt[n]{a^m}} = \frac{1}{\left(\sqrt[n]{a}\right)^m}$$

Thus:

$$8^{-\frac{1}{3}} = \frac{1}{\sqrt[3]{8}} = \frac{1}{2}$$

All of these relationships are very useful when simplifying complicated radical and exponential expressions. If an expression contains both forms, use one of these rules to change the expression to contain either all radicals or all exponential expressions. This process makes the entire expression much easier to work with, especially if the expressions are contained within equations.

Consider the following example:

$$\sqrt{x} \times \sqrt[4]{x}$$

It is written in radical form; however, it can be simplified into one radical by using exponential expressions first. The expression can be written as:

$$x^{\frac{1}{2}} \times x^{\frac{1}{4}}$$

It can be combined into one base by adding the exponents as:

$$x^{\frac{1}{2}+\frac{1}{4}} = x^{\frac{3}{4}}$$

Writing this back in radical form, the result is:

$$\sqrt[4]{x^3}$$

Systems of Equations

Creating, Solving, or Interpreting Systems of Linear Inequalities in Two Variables

A **system of linear inequalities in two variables** consists of two inequalities in two variables, typically x and y. For example, the following is a system of linear inequalities in two variables:

$$\begin{cases} 4x + 2y < 1 \\ 2x - y \leq 0 \end{cases}$$

The curly brace on the left side shows that the two inequalities are grouped together. A solution of a single inequality in two variables is an ordered pair that satisfies the inequality. For example, (1, 3) is a solution of the linear inequality $y \geq x + 1$ because when plugged in, it results in a true statement. The graph of an inequality in two variables consists of all ordered pairs that make the solution true. Therefore, the entire solution set of a single inequality contains many ordered pairs, and the set can be graphed by using a half plane.

A **half plane** consists of the set of all points on one side of a line. If the inequality consists of $>$ or $<$, the line is dashed because no solutions actually exist on the line shown. If the inequality consists of $\geq$ or $\leq$, the line is solid and solutions are on the line shown. To graph a linear inequality, graph the corresponding equation found by replacing the inequality symbol with an equals sign. Then pick a test point that exists on either side of the line. If that point results in a true statement when plugged into the original inequality, shade in the side containing the test point. If it results in a false statement, shade in the opposite side.

Solving a system of linear inequalities must be done graphically. Follow the process as described above for both given inequalities. The solution set to the entire system is the region that is in common to every graph in the system. For example, here is the solution to the following system:

$$\begin{cases} y \geq 3 - x \\ y \leq -3 - x \end{cases}$$

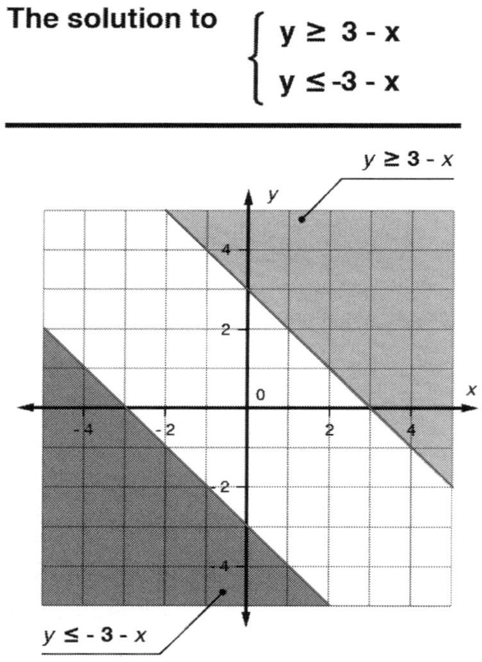

Note that there is no region in common, so this system has no solution.

Creating, Solving, or Interpreting Systems of Two Linear Equations in Two Variables
An example of a system of two linear equations in two variables is the following:

$$2x + 5y = 8$$

$$5x + 48y = 9$$

A solution to a **system of two linear equations** is an ordered pair that satisfies both the equations in the system. A system can have one solution, no solution, or infinitely many solutions. The solution can be found through a graphing technique. The solution to a system of equations is actually equal to the point where both lines intersects. If the lines intersect at one point, there is one solution and the system is said to be **consistent**. However, if the two lines are parallel, they will never intersect and there is no solution. In this case, the system is said to be **inconsistent**. Thirdly, if the two lines are actually the same line, there are infinitely many solutions and the solution set is equal to the entire line. The lines are **dependent**.

Here is a summary of the three cases:

Solving Systems by Graphing

Consistent	Inconsistent	Dependent
One solution	No solution	Infinite number of solutions
Lines intersect	*Lines are parallel*	*Coincide: same line*

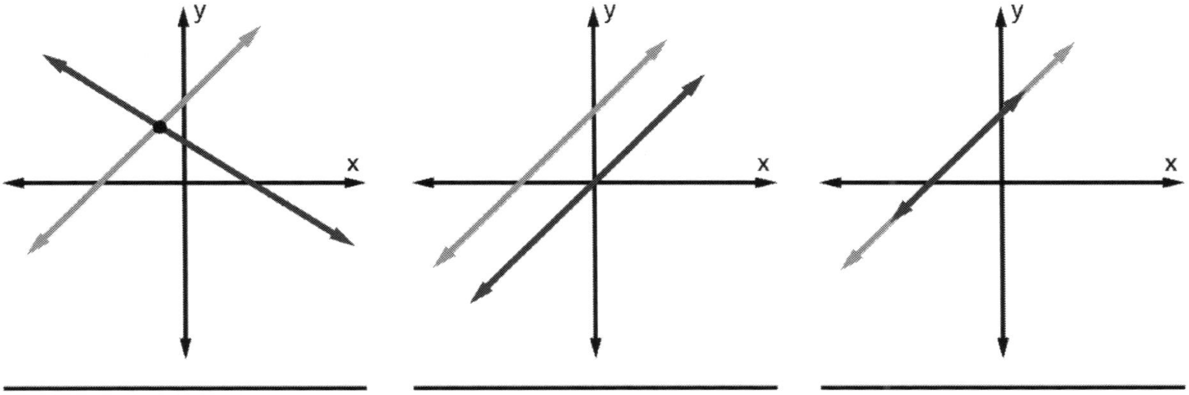

Consider the following system of equations:

$$\begin{cases} y + x = 3 \\ y - x = 1 \end{cases}$$

To find the solution graphically, graph both lines on the same *xy*-plane. Graph each line using either a table of ordered pairs, the *x*- and *y*-intercepts, or slope and the *y*-intercept. Then, locate the point of intersection.

The graph is shown here:

The System of Equations $\begin{cases} y + x = 3 \\ y - x = 1 \end{cases}$

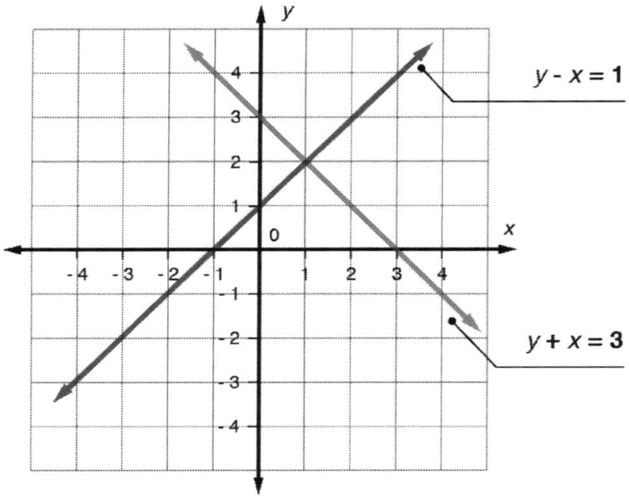

It can be seen that the point of intersection is the ordered pair (1, 2). This solution can be checked by plugging it back into both original equations to make sure it results in true statements. This process results in:

$$2 + 1 = 3$$

$$2 - 1 = 1$$

Both equations are true, so the solution is correct.

The following system has no solution:

$$y = 4x + 1$$

$$y = 4x - 1$$

Both lines have the same slope and different y-intercepts, so they are parallel, meaning that they run alongside each other and never intersect.

Finally, the following solution has infinitely many solutions:

$$2x - 7y = 12$$

$$4x - 14y = 24$$

Note that the second equation is equal to the first equation times 2. Therefore, they are the same line. The solution set can be written in set notation as $\{(x, y)|2x - 7y = 12\}$, which represents the entire line.

Algebraically Solving Systems of Two Linear Equations in Two Variables

There are two algebraic methods to finding solutions. The first is **substitution**. This process is better suited for systems when one of the equations is already solved for one variable, or when solving for one variable is easy to do. The equation that is already solved for is substituted into the other equation for that variable, and this process results in a linear equation in one variable. This equation can be solved for the given variable, and then that solution can be plugged into one of the original equations, which can then be solved for the other variable. This last step is known as **back-substitution**, and the end result is an ordered pair.

The following is an example of a system that is suited for substitution:

$$y = 4x + 2$$

$$2x + 3y = 9$$

The other method is known as **elimination,** or the **addition method**. This is better suited when the equations are in standard form:

$$Ax + By = C$$

The goal in this method is to multiply one or both equations times numbers that result in opposite coefficients. Then, add the equations together to obtain an equation in one variable. Solve for the given variable, then take that value and back-substitute to obtain the other part of the ordered pair solution.

The following is an example of a system that is suited for elimination:

$$2x + 3y = 8$$

$$4x - 2y = 10$$

Note that in order to check an answer when solving a system of equations, the solution must be checked in both original equations to show that it solves not only one of the equations, but both of them.

If either solution results in an untrue statement when inserted into the original equation, then there is no solution to the system. Finally, if throughout either solution procedure the process results in the variables dropping out, which gives a statement that is always true, there are infinitely many solutions.

Functions

Function Definition

A **relation** is any set of ordered pairs (x, y). The values listed first in the ordered pairs, known as the x-coordinates, make up the domain of the relation. The values listed second, known as the y-coordinates, make up the range. A relation in which every member of the domain corresponds to only one member of the range is known as a **function**. A function cannot have a member of the domain corresponding to two members of the range.

Function Notation

Functions are most often given in terms of equations instead of ordered pairs. For instance, here is an equation of a line:

$$y = 2x + 4$$

In function notation, this can be written as:

$$f(x) = 2x + 4$$

The expression $f(x)$ is read "f of x" and it shows that the inputs, the x-values, get plugged into the function and the output is:

$$y = f(x)$$

The set of all inputs are in the domain, and the set of all outputs are in the range.

The x-values are known as the **independent variables** of the function and the y-values are known as the **dependent variables** of the function. The y-values depend on the x-values. For instance, if $x = 2$ is plugged into the function shown above, the y-value depends on that input.

$$f(2) = 2 \times 2 + 4 = 8$$

Therefore, $f(2) = 8$, which is the same as writing the ordered pair (2, 8). To graph a function, graph it in equation form. Therefore, replace $f(x)$ with y and plot ordered pairs.

Due to the definition of a function, the graph of a function cannot have two of the same x-components paired to different y-component. For example, the ordered pairs (3, 4) and (3, -1) cannot be in a valid function. Therefore, all graphs of functions pass the **vertical line test**. If any vertical line intersects a graph in more than one place, the graph is not that of a function. For instance, the graph of a circle is not a function because one can draw a vertical line through a circle and intersect the circle twice. Common functions include lines and polynomials, which pass the vertical line test.

Representing Functions

When presented with a real-world problem, the first step is to determine what unknown quantity must be solved for. Use a variable, such as x or t, to represent that unknown quantity. Sometimes, there can be two or more unknown quantities. In this case, either choose an additional variable, or if a relationship exists between the unknown quantities, express the other quantities in terms of the original variable. After choosing the variables, form algebraic expressions and/or equations that represent the verbal statement in the problem. The following table shows examples of vocabulary used to represent the different operations:

Addition	Sum, plus, total, increase, more than, combined, in all
Subtraction	Difference, less than, subtract, reduce, decrease, fewer, remain
Multiplication	Product, multiply, times, part of, twice, triple
Division	Quotient, divide, split, each, equal parts, per, average, shared

The combination of operations and variables form both mathematical expression and equations. The differences between expressions and equations are that there is no equals sign in an expression, and that expressions are evaluated to find an unknown quantity, while equations are solved to find an unknown quantity. Also, inequalities can exist within verbal mathematical statements. Instead of a statement of equality, expressions state quantities are *less than*, *less than or equal to*, *greater than*, or *greater than or equal to*. Another type of inequality is when a quantity is said to be not equal to another quantity ($\neq$).

The steps for solving inequalities in one variable are the same steps for solving equations in one variable. The addition and multiplication principles are used. However, to maintain a true statement when using the $<, \leq, >$, and $\geq$ symbols, if a negative number is either multiplied by both sides of an inequality or divided from both sides of an inequality, the sign must be flipped. For instance, consider the following inequality:

$$3 - 5x \leq 8$$

First, 3 is subtracted from each side to obtain:

$$-5x \leq 5$$

Then, both sides are divided by -5, while flipping the sign, to obtain:

$$x \geq -1$$

Therefore, any real number greater than or equal to -1 satisfies the original inequality.

Linear Functions

A linear function that models a linear relationship between two quantities is of the form $y = mx + b$, or in function form $f(x) = mx + b$. In a linear function, the value of y depends on the value of x, and y increases or decreases at a constant rate as x increases. Therefore, the independent variable is x, and the dependent variable is y. The graph of a linear function is a line, and the constant rate can be seen by looking at the steepness, or slope, of the line. If the line increases from left to right, the slope is positive. If the line slopes downward from left to right, the slope is negative.

In the function, m represents slope. Each point on the line is an **ordered pair** (x, y), where x represents the x-coordinate of the point and y represents the y-coordinate of the point. The point where $x = 0$ is known as the y-intercept, and it is the place where the line crosses the y-axis. If $x = 0$ is plugged into $f(x) = mx + b$, the result is $f(0) = b$; therefore, the point $(0, b)$ is the y-intercept of the line. The derivative of a linear function is its slope.

Consider the following situation. A taxicab driver charges a flat fee of \$2 per ride and \$3 a mile. This statement can be modeled by the function $f(x) = 3x + 2$ where x represents the number of miles and $f(x) = y$ represents the total cost of the ride. The total cost increases at a constant rate of \$2 per mile, and that is why this situation is a linear relationship. The slope $m = 3$ is equivalent to this rate of change. The flat fee of \$2 is the y-intercept. It is the place where the graph crosses the x-axis, and it represents the cost when $x = 0$, or when no miles have been traveled in the cab. The y-intercept in this situation represents the flat fee.

A linear function of the form $f(x) = mx + b$ has two important quantities: m and b. The quantity m represents the slope of the line, and the quantity b represents the y-intercept of the line. When the function represents a real-life situation or a mathematical model, these two quantities are very meaningful. The slope, m, represents the rate of change, or the amount y increases or decreases given an increase in x. If m is positive, the rate of change is positive, and if m is negative, the rate of change is negative. The y-intercept, b, represents the amount of quantity y when x is 0. In many applications, if the x-variable is never a negative quantity, the y-intercept represents the initial amount of the quantity y. The x-variable often represents time, so it makes sense that it would not be negative.

Consider the following example. These two equations represent the cost, *C*, of t-shirts, *x*, at two different printing companies:

$$C(x) = 7x$$

$$C(x) = 5x + 25$$

The first equation represents a scenario in which each t-shirt costs \$7. In this equation, *x* varies directly with *y*. There is no *y*-intercept, which means that there is no initial cost for using that printing company. The rate of change is 7, which is price per shirt. The second equation represents a scenario that has both an initial cost and a cost per t-shirt. The slope 5 shows that each shirt is \$5. The *y*-intercept 25 shows that there is an initial cost of \$25 when using that company. Therefore, it makes sense to use the first company at \$7 per shirt when only purchasing a small number of t-shirts. However, large orders would be cheaper by going with the second company because eventually that initial cost will become negligible.

Radical Functions

Recall that a **radical expression** is an expression involving a square root, a cube root, or a higher order root such as fourth root, fifth root, etc. The expression underneath the radical is known as the **radicand** and the index is the number corresponding to the **root**. An index of 2 corresponds to a square root. A radical function is a function that involves a radical expression. For instance, $\sqrt{x+1}$ is a radical expression, $x + 1$ is the radicand, and the corresponding function is:

$$y = \sqrt{x+1}$$

It can also be written in function notation as:

$$f(x) = \sqrt{x+1}$$

If the root is even, meaning a square root, fourth root, etc., the radicand must be positive. Therefore, in order to find the domain of a radical function with an even index, set the radicand greater than or equal to zero and find the set of numbers that satisfies that inequality. The domain of $f(x) = \sqrt{x+1}$ is all numbers greater than or equal to -1. The range of this function is all nonnegative real numbers because the square root, or any even root, can never output a negative number. The domain of an odd root is all real numbers because the radicand can be negative in an odd root.

Piecewise Functions

A **piecewise function** is basically a function that is defined in pieces. The graph of the function behaves differently over different intervals along the *x*-axis, or different intervals of its domain. Therefore, the function is defined using different mathematical expressions over these intervals. The function is not defined by only one equation. In a piecewise function, the function is actually defined by two or more equations, where each equation is used over a specific interval of the domain.

Here is a graph of a piecewise function:

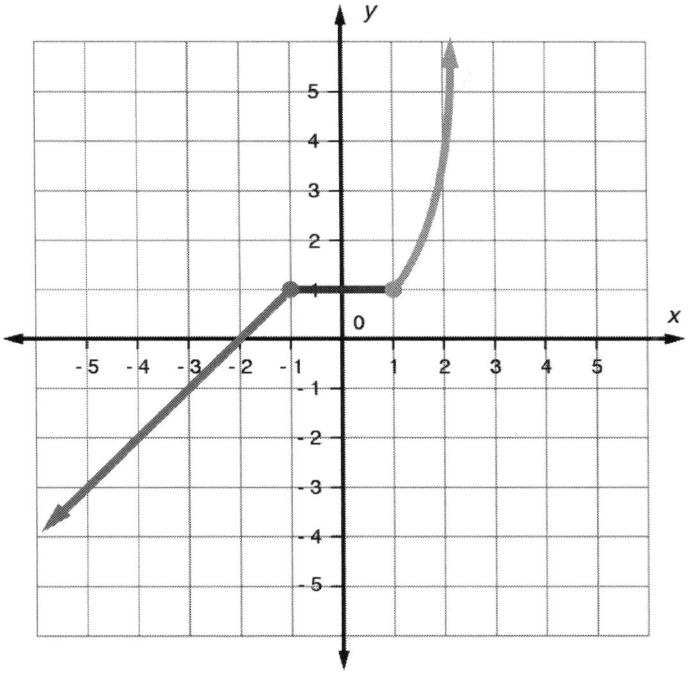

Notice that from $(-\infty, -1]$, the graph is a line with positive slope. From $[-1, 1]$ the graph is a horizontal line. Finally, from $[1, \infty)$ the graph is a nonlinear curve. Both the domain and range of this function are all real numbers, expressed as $(-\infty, \infty)$.

Piecewise functions can also have **discontinuities**, which are jumps in the graph. When drawing a graph, if the pencil must be picked up at any point to continue drawing, the graph has a discontinuity. Here is the graph of a piecewise function with discontinuities at $x = 1$ and $x = 2$:

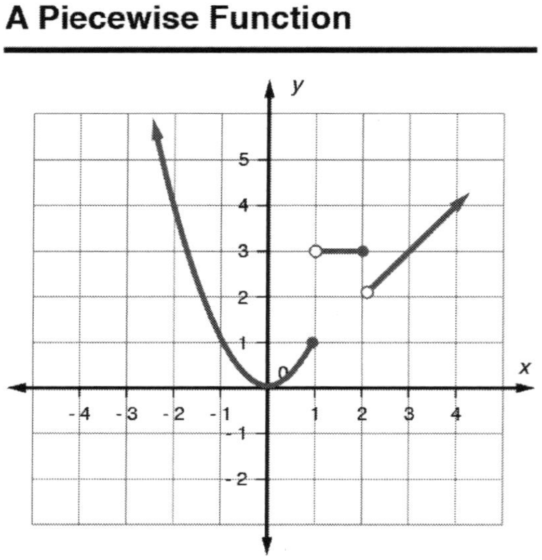

The open circle at a point indicates that the endpoint is not included in that part of the graph, and the closed circle indicates that the endpoint is included. The domain of this function is all real numbers; however, the range is all nonnegative real numbers $[0, \infty)$.

Polynomial Functions

A **polynomial function** is a function containing a polynomial expression, which is an expression containing constants and variables combined using the four mathematical operations. The degree of a polynomial depends on the largest exponent in the expression. Typical polynomial functions are **quartic,** with a degree of 4, **cubic,** with a degree of 3, and **quadratic,** with a degree of 2. Note that the exponents on the variables can only be nonnegative integers. The domain of any polynomial function is all real numbers because any number plugged into a polynomial expression grants a real number output. An example of a quartic polynomial equation is:

$$y = x^4 + 3x^3 - 2x + 1$$

The zeros of a polynomial function are the points where its graph crosses the y-axis. In order to find the number of real zeros of a polynomial function, use **Descartes' Rule of Signs**, which states that the number of possible positive real zeros is equal to the number of sign changes in the coefficients. If there is only one sign change, there is only one positive real zero. In the example above, the signs of the coefficients are positive, positive, negative, and positive. The sign changes twice; therefore, there are at most two positive real zeros. The number of possible negative real zeros is equal to the number of sign changes in the coefficients when plugging $-x$ into the equation. Again, if there is only one sign change, there is only one negative real zero. The polynomial result when plugging $-x$ into the equation is:

$$y = (-x)^4 + 3(-x)^3 - 2(-x) + 1$$

$$y = x^4 - 3x^3 + 2x + 1$$

The sign changes two times, so there are at most two negative real zeros. Another polynomial equation this rule can be applied to is:

$$y = x^3 + 2x - x - 5$$

There is only one sign change in the terms of the polynomial, so there is exactly one real zero. When plugging $-x$ into the equation, the polynomial result is:

$$-x^3 - 2x - x - 5$$

There are no sign changes in this polynomial, so there are no possible negative zeros.

Logarithmic Functions

For $x > 0, b > 0, b \neq 1$, the function $f(x) = \log_b x$ is known as the **logarithmic function** with base *b*. With $y = \log_b x$, its exponential equivalent is $b^y = x$. In either case, the exponent is *y* and the base is *b*. Therefore, $3 = \log_2 8$ is the same as $2^2 = 8$. So, in order to find the logarithm with base 2 of 8, find the exponent that when 2 is raised to that value results in 8. Similarly:

$$\log_3 243 = 5$$

In order to do this mentally, ask the question, what exponent does 3 need to be raised to that results in 243? The answer is 5. Most logarithms do not have whole number results. In this case, a calculator can be used. A calculator typically has buttons with base 10 and base e, so the change of base formula can be used to calculate these logs.

For instance:

$$\log_3 55 = \frac{\log 55}{\log 3} = 3.64$$

Similarly, the natural logarithm with base e could be used to obtain the same result.

$$\log_3 55 = \frac{\ln 55}{\ln 3} = 3.64$$

The domain of a logarithmic function $f(x) = \log_b x$ is all positive real numbers. This is because the exponent must be a positive number. The range of a logarithmic function $f(x) = \log_b x$ is all real numbers. The graphs of all logarithmic functions of the form $f(x) = \log_b x$ always pass through the point (1, 0) because anything raised to the power of 0 is 1. Therefore, such a function always has an *x*-intercept at 1. If the base is greater than 1, the graph increases from the left to the right along the *x*-axis. If the base is between 0 and 1, the graph decreases from the left to the right along the *x*-axis. In both situations, the *y*-axis is a vertical **asymptote**. The graph will never touch the *y*-axis, but it does approach it closely.

Here are the graphs of the two cases of logarithmic functions:

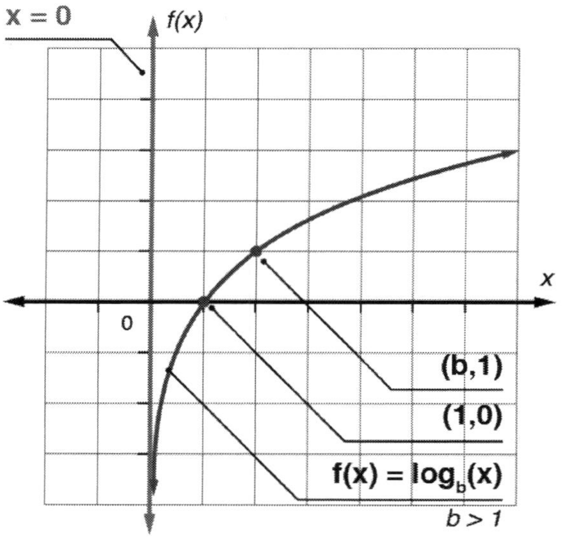

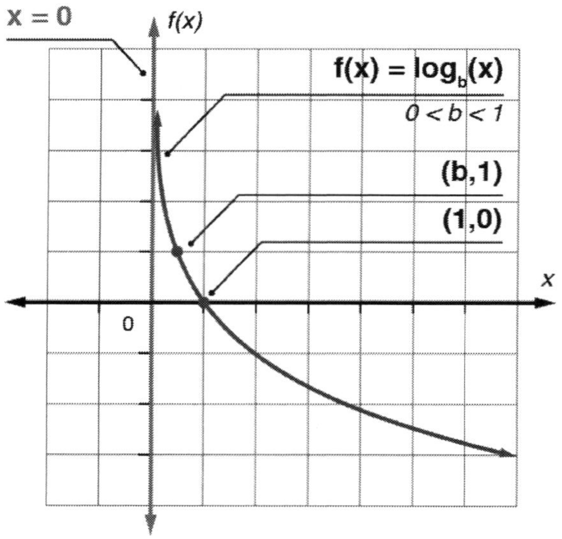

Finding and Applying Important Features of Graphs

A **graph** is a pictorial representation of the relationship between two variables. To read and interpret a graph, it is necessary to identify important features of the graph. First, read the title to determine what data sets are being related in the graph. Next, read the axis labels and understand the scale that is used. The horizontal axis often displays categories, like years, month, or types of pets. The vertical axis often displays numerical data like amount of income, number of items sold, or number of pets owned. Check to see what increments are used on each axis. The changes on the axis may represent fives, tens, hundreds, or any increment. Be sure to note what the increment is because it will affect the interpretation of the graph. Now, locate on the graph an element of interest and move across to find the element to which it relates.

For example, notice an element displayed on the horizontal axis, find that element on the graph, and then follow it across to the corresponding point on the vertical axis. Using the appropriate scale, interpret the relationship.

Choosing Appropriate Graphs to Display Data

Data may be displayed with a line graph, bar graph, or pie chart.

- A line graph is used to display data that changes continuously over time.
- A bar graph is used to compare data from different categories or groups and is helpful for recognizing relationships.
- A pie chart is used when the data represents parts of a whole.

Identifying, Constructing, and Completing Graphs That Correctly Represent Given Data

Data is often displayed with a line graph, bar graph, or pie chart.

The line graph below shows the number of push-ups that a student did over one week.

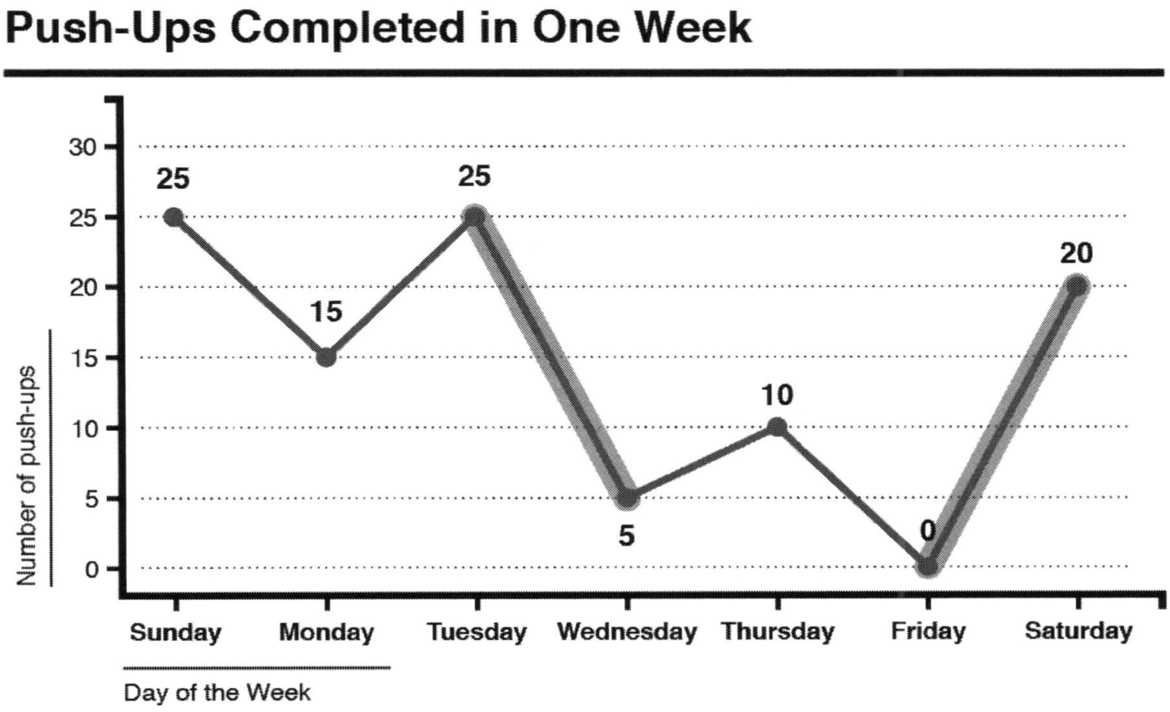

Notice that the horizontal axis displays the day of the week and the vertical axis displays the number of push-ups. A point is placed above each day of the week to show how many push-ups were done each day. For example, on Sunday the student did 25 push-ups. The line that connects the points shows how much the number of push-ups fluctuated throughout the week.

The bar graph below compares number of people who own various types of pets.

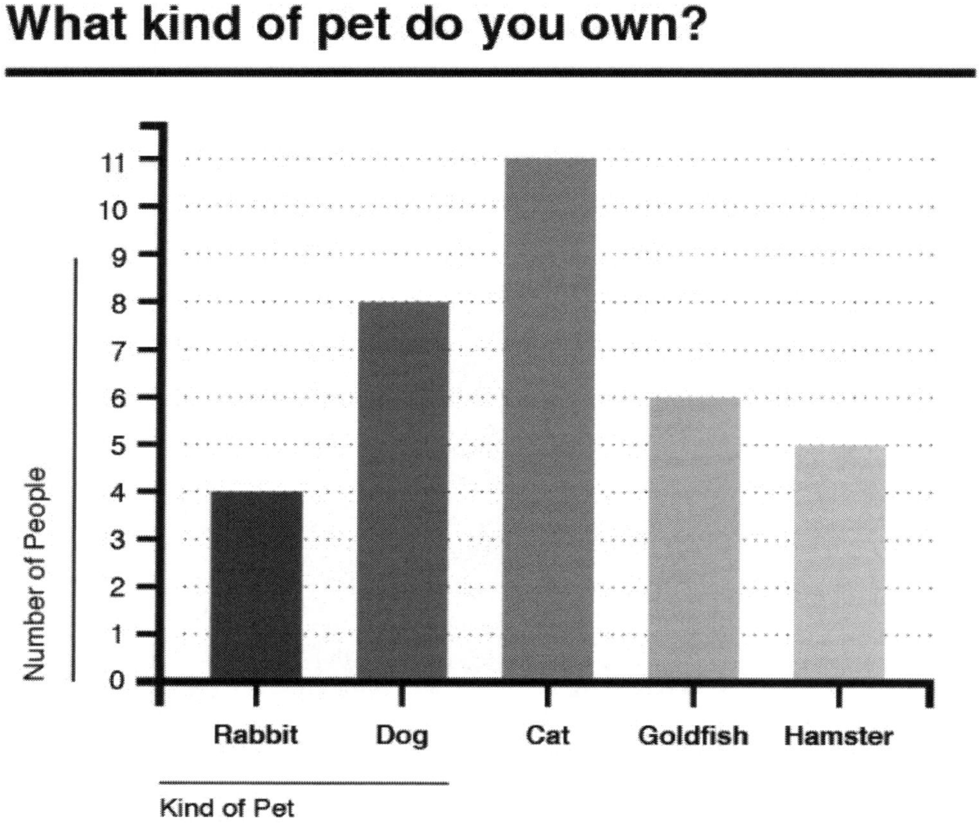

On the horizontal axis, the kind of pet is displayed. On the vertical axis, the number of people is displayed. Bars are drawn to show the number of people who own each type of pet. With the bar graph, it can quickly be determined that the fewest number of people own a rabbit and the greatest number of people own a cat.

The pie graph below displays students in a class who scored A, B, C, or D. Each slice of the pie is drawn to show the portion of the whole class that is represented by each letter grade. For example, the smallest portion represents students who scored a D. This means that the fewest number of students scored a D.

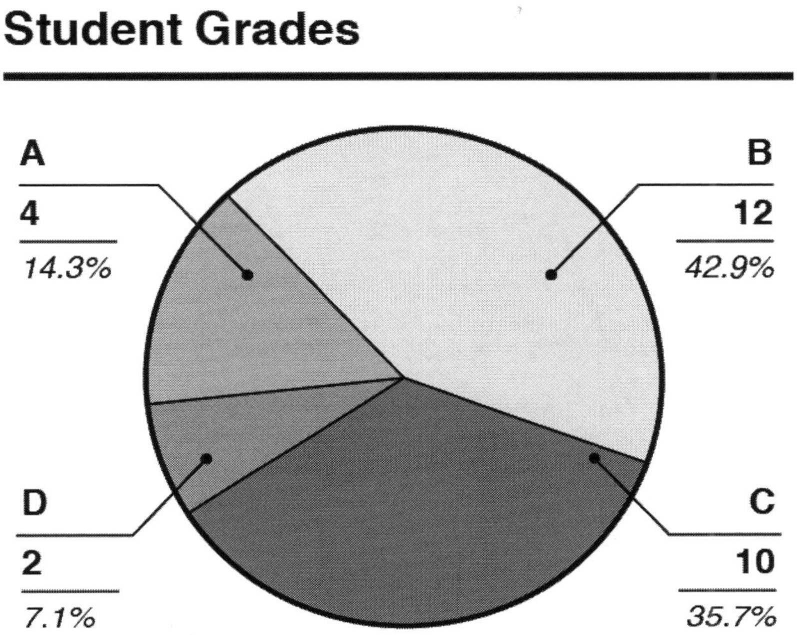

Geometry

Defining and Applying Knowledge of Shapes and Solids

Shapes are defined by their angles and number of sides. A shape with one continuous side, where all points on that side are equidistant from a center point is called a **circle.** A shape made with three straight line segments is a **triangle.** A shape with four sides is called a **quadrilateral,** but more specifically a **square, rectangle, parallelogram**, or **trapezoid,** depending on the interior angles. These shapes are two-dimensional and only made of straight lines and angles. **Solids** can be formed by combining these shapes and forming three-dimensional figures. While two-dimensional figures have only length and height, three-dimensional figures also have depth. Examples of solids include prisms or spheres.

The four figures below have different names based on their sides and dimensions. Figure 1 is a **cone** because it has three dimensions, where the bottom is a circle and the top is formed by the sides combining to one point. Figure 2 is a **triangle** because it has two dimensions, made up of three-line segments.

Figure 3 is a **cylinder** made up of two base circles and a rectangle to connect them in three dimensions. Figure 4 is an **oval** formed by one continuous line in two dimensions; it differs from a circle because not all points are equidistant from the center.

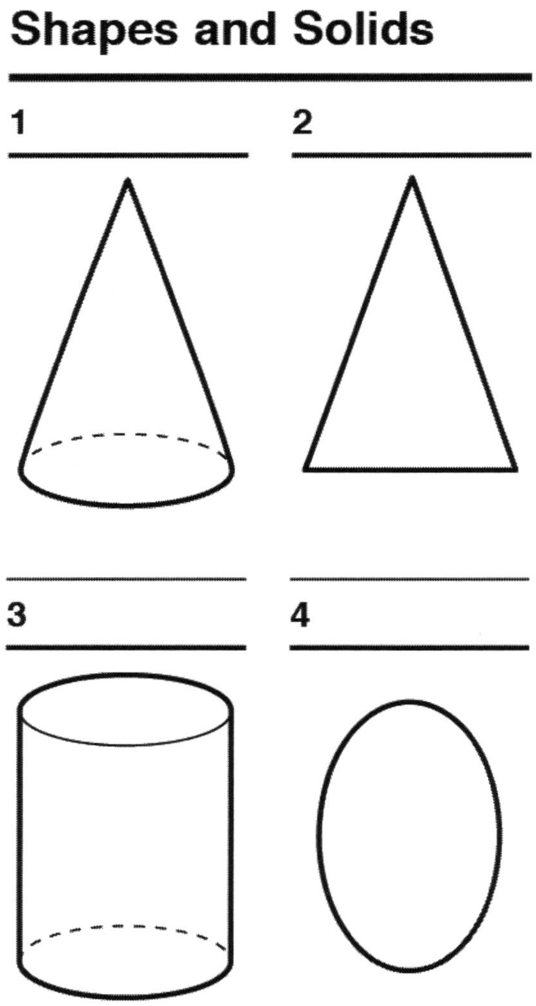

The **cube** in Figure 5 below is a three-dimensional solid made up of squares. Figure 6 is a **rectangle** because it has four sides that intersect at right angles. More specifically, it is a square because the four sides are equal in length. Figure 7 is a **pyramid** because the bottom shape is a square and the sides are all triangles. These triangles intersect at a point above the square. Figure 8 is a **circle** because it is made up of one continuous line where the points are all equidistant from one center point.

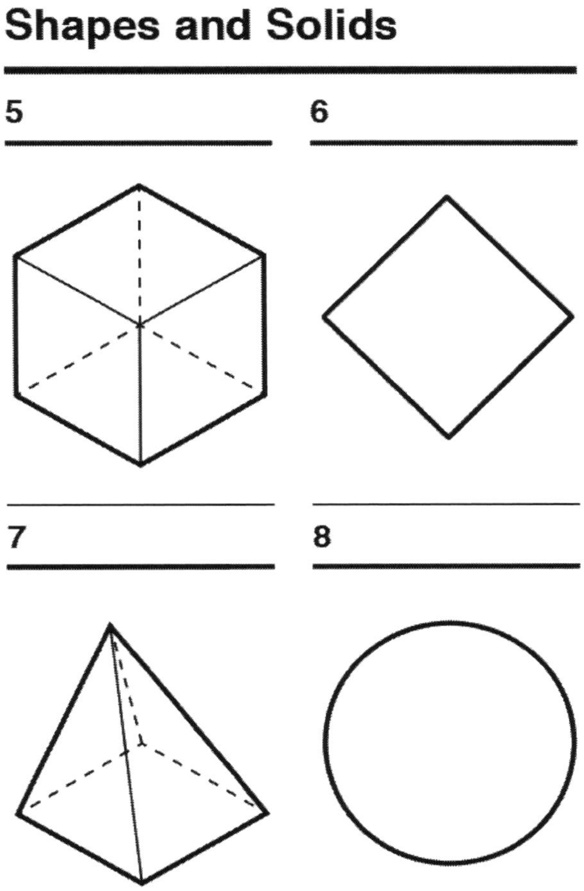

Congruence and Similarity

Two figures are **congruent** if they have the same shape and same size, meaning same angle measurements and equal side lengths. Two figures are similar if they have the same angle measurements but not side lengths. Basically, angles are congruent in similar triangles and their side lengths are constant multiples of each other. Proving two shapes are similar involves showing that all angles are the same; proving two shapes are congruent involves showing that all angles are the same and that all sides are the same. If two pairs of angles are congruent in two triangles, then those triangles are similar because their third angles have to be equal due to the fact that all three angles add up to 180 degrees.

There are five main theorems that are used to prove congruence in triangles. Each theorem involves showing that different combinations of sides and angles are the same in two triangles, which proves the triangles are congruent. The **side-side-side (SSS) theorem** states that if all sides are equal in two triangles, the triangles are congruent. The **side-angle-side (SAS) theorem** states that if two pairs of sides and the included angles are equal in two triangles then the triangles are congruent. Similarly, the **angle-**

side-angle (ASA) theorem states that if two pairs of angles and the included side lengths are equal in two triangles, the triangles are similar. The **angle-angle-side (AAS) theorem** states that two triangles are congruent if they have two pairs of congruent angles and a pair of corresponding equal side lengths that are not included. Finally, the **hypotenuse-leg (HL) theorem** states that if two right triangles have equal hypotenuses and an equal pair of shorter sides, the triangles are congruent. An important item to note is that angle-angle-angle (AAA) is not enough information to prove congruence because the three angles could be equal in two triangles, but their sides could be different lengths.

Using the Relationship Between Similarity, Right Triangles, and Trigonometric Ratios
Within two similar triangles, corresponding side lengths are proportional, and angles are equal. In other words, regarding corresponding sides in two similar triangles, the ratio of side lengths is the same. Recall that the SAS theorem for similarity states that if an angle in one triangle is congruent to an angle in a second triangle, and the lengths of the sides in both triangles are proportional, then the triangles are similar. Also, because the ratio of two sides in two similar right triangles is the same, the trigonometric ratios in similar right triangles are always going to be equal.

If two triangles are similar, and one is a right triangle, the other is a right triangle. The definition of similarity ensures that each triangle has a 90-degree angle. In a similar sense, if two triangles are right triangles containing a pair of equal acute angles, the triangles are similar because the third pair of angles must be equal as well. However, right triangles are not necessarily always similar.

The following triangles are similar:

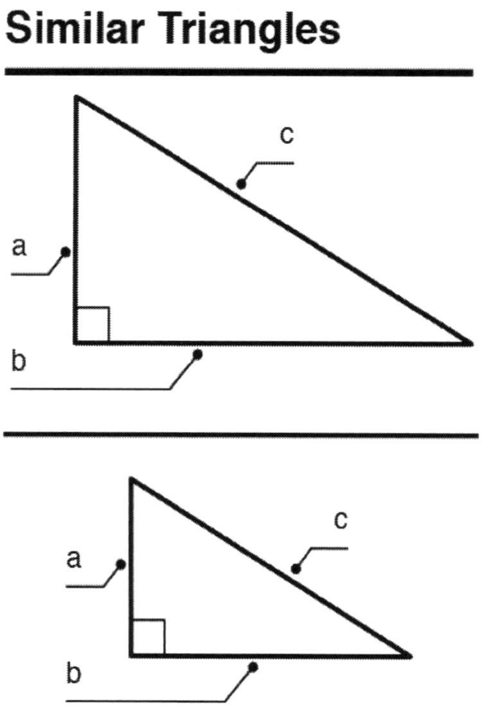

It is not always apparent at first glance, but theorems can be used to show similarity. The **Pythagorean Theorem** can be used to find the missing side lengths in both triangles. In the larger triangle, the missing side is the hypotenuse, *c*. Therefore:

$$9^2 + 12^2 = c^2$$

This equation is equivalent to $225 = c^2$, so taking the square root of both sides results in the positive root $c = 15$. In the other triangle, the Pythagorean Theorem can be used to find the missing side length *b*. The theorem shows that $6^2 + b^2 = 10^2$, and *b* is then solved for to obtain $b = 8$. The ratio of the sides in the larger triangle to the sides in the smaller triangle is the same value, 1.5. Therefore, the sides are proportional. Because they are both right triangles, they have a congruent angle. The SAS theorem for similarity can be used to show that these two triangles are similar.

Surface Area and Volume Measurements

Surface area is defined as the area of the surface of a figure. A **pyramid** has a surface made up of four triangles and one square. To calculate the surface area of a pyramid, the areas of each individual shape are calculated. Then the areas are added together. This method of decomposing the shape into two-dimensional figures to find area, then adding the areas, can be used to find surface area for any figure. Once these measurements are found, the area is described with square units. For example, the following figure shows a rectangular prism. The figure beside it shows the rectangular prism broken down into two-dimensional shapes, or rectangles. The area of each rectangle can be calculated by multiplying the length by the width. The area for the six rectangles can be represented by the following expression:

$$5 \times 6 + 5 \times 10 + 5 \times 6 + 6 \times 10 + 5 \times 10 + 6 \times 10$$

The total for all these areas added together is 280 m^2, or 280 square meters.

This measurement represents the surface area because it is the area of all six surfaces of the rectangular prism.

The Net of a Rectangular Prism

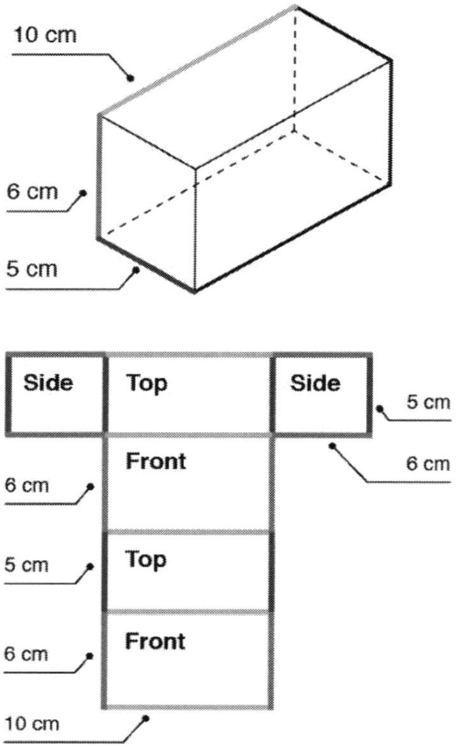

The surface area of a three-dimensional figure is the total area of each of the figure's faces. Because nets lay out each face of an object, they make it easier to visualize and measure surface area. The following figure shows a triangular prism. The bases are triangles and the sides are rectangles. The second figure shows the net for this triangular prism. The dimensions are labeled for each of the faces of the prism. To determine the area for the two triangles, use the following formula:

$$A = \frac{1}{2}bh$$

$$A = \frac{1}{2} \times 8 \times 9 = 36 \text{ cm}^2$$

The rectangles' areas can be described by the equation:

$$A = lw = 8 \times 5 + 9 \times 5 + 10 \times 5$$

$$A = 40 + 45 + 50 = 135 \text{ cm}^2$$

The area for the triangles can be multiplied by two, then added to the rectangle areas to yield a total surface area of $207\ \text{cm}^2$.

A Triangular Prism and Its Net

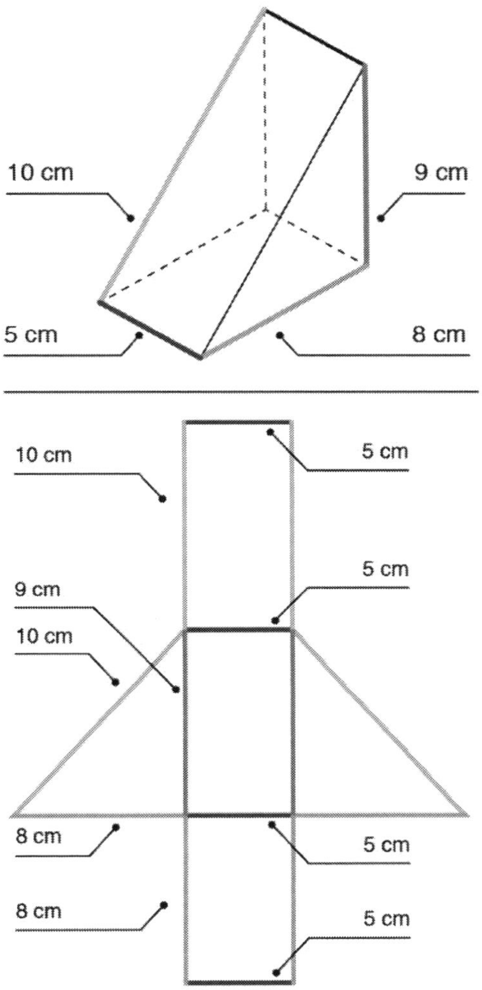

$$\text{SA} = 2 \times \left(\frac{1}{2}\text{bh}\right) + \text{lw}$$
$$= 2 \times \left(\frac{1}{2} \times 8 \times 9\right) + (8 \times 5 + 9 \times 5 + 10 \times 5)$$
$$= 207\text{cm}^2$$

Another shape that has a surface area is a cylinder. The shapes used to make up the **cylinder** are two circles and a rectangle wrapped around between the two circles. A common example of a cylinder is a can. The two circles that make up the bases are obvious shapes. The rectangle can be more difficult to see, but the label on a can will help illustrate it. When the label is removed from a can and laid flat, the shape is a rectangle.

When the areas for each shape are needed, there will be two formulas. The first is the area for the circles on the bases. This area is given by the formula $A = \pi r^2$. There will be two of these areas. Then the area of the rectangle must be determined. The width of the rectangle is equal to the height of the can, h. The length of the rectangle is equal to the circumference of the base circle, $2\pi r$. The area for the rectangle can be found by using the formula:

$$A = 2\pi r \times \mathrm{h}$$

By adding the two areas for the bases and the area of the rectangle, the surface area of the cylinder can be found, described in units squared.

Finding the Volume and Surface Area of Right Rectangular Prisms, Including Those with Fractional Edge Lengths

Right rectangular prisms are those prisms in which all sides are rectangles, and all angles are right, or equal, to 90 degrees. The volume for these objects can be found by multiplying the length by the width by the height. The formula is $V = lwh$. For the following prism, the volume formula is:

$$V = 6\frac{1}{2} \times 3 \times 9$$

When dealing with fractional edge lengths, it is helpful to convert the length to an improper fraction. The length $6\frac{1}{2}$ cm becomes $\frac{13}{2}$ cm. Then the formula becomes:

$$V = \frac{13}{2} \times 3 \times 9$$

$$\frac{13}{2} \times \frac{3}{1} \times \frac{9}{1} = \frac{351}{2}$$

This value for volume is better understood when turned into a mixed number, which would be $175\frac{1}{2}\ \mathrm{cm}^3$.

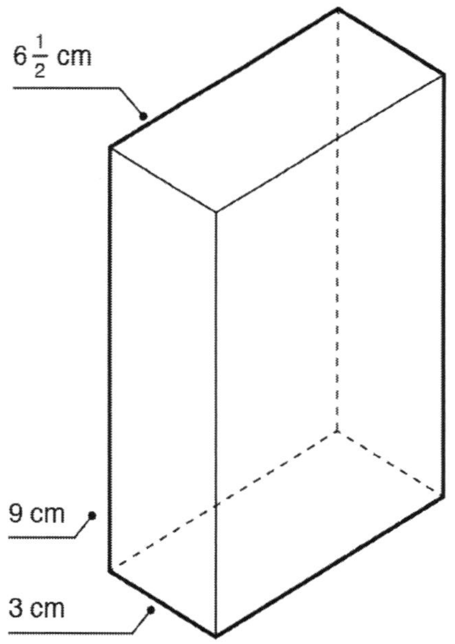

When dimensions for length are given with fractional parts, it can be helpful to turn the mixed number into an improper fraction, then multiply to find the volume, then convert back to a mixed number. When finding surface area, this conversion to improper fractions can also be helpful. The surface area can be found for the same prism above by breaking down the figure into basic shapes. These shapes are rectangles, made up of the two bases, two sides, and the front and back.

The formula for surface area adds the areas for each of these shapes in the following equation:

$$SA = 6\frac{1}{2} \times 3 + 6\frac{1}{2} \times 3 + 3 \times 9 + 3 \times 9 + 6\frac{1}{2} \times 9 + 6\frac{1}{2} \times 9$$

Because there are so many terms in a surface area formula and because this formula contains a fraction, it can be simplified by combining groups that are the same. Each set of numbers is used twice, to represent areas for the opposite sides of the prism. The formula can be simplified to:

$$SA = 2\left(6\frac{1}{2} \times 3\right) + 2(3 \times 9) + 2\left(6\frac{1}{2} \times 9\right)$$

$$2\left(\frac{13}{2} \times 3\right) + 2(27) + 2\left(\frac{13}{2} \times 9\right)$$

$$2\left(\frac{39}{2}\right) + 54 + 2\left(\frac{117}{2}\right)$$

$$39 + 54 + 117 = 210 \text{ cm}^2$$

Understanding Composition of Objects

Composition of objects is the way objects are used in conjunction with each other to form bigger, more complex shapes. For example, a rectangle and a triangle can be used together to form an arrow. Arrows can be found in many everyday scenarios, but are often not seen as the composition of two different shapes. A square is a common shape, but it can also be the composition of shapes. As seen in the second figure, there are many shapes used in the making of the one square. There are five triangles that are three different sizes. There is also one square and one parallelogram used to compose this square. These shapes can be used to compose each more complex shape because they line up, side by side, to fill in the shape with no gaps. This defines composition of shapes where smaller shapes are used to make larger, more complex ones.

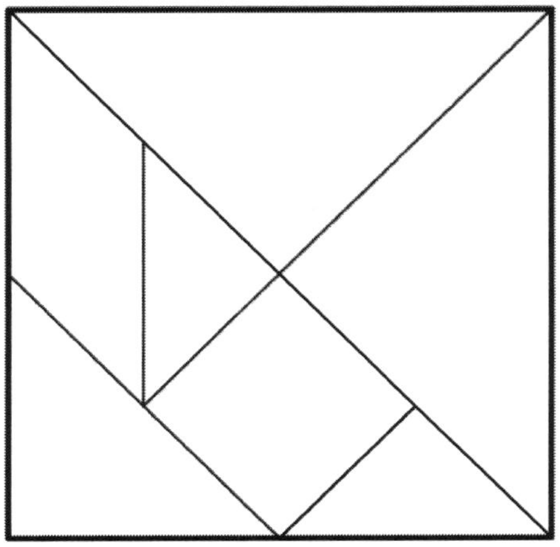

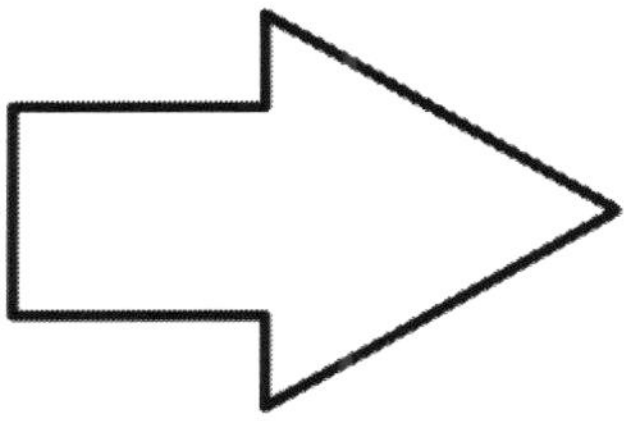

Solving for Missing Values in Triangles, Circles, and Other Figures

Solving for missing values in shapes requires knowledge of the shape and its characteristics. For example, a triangle has three sides and three angles that add up to 180 degrees. If two angle measurements are given, the third can be calculated. For the triangle below, the one given angle has a measure of 55 degrees. The missing angle is x. The third angle is labeled with a square, which indicates a measure of 90 degrees. Because all angles must add up to 180 degrees,, the following equation can be used to find the missing x-value:

$$55° + 90° + x = 180°$$

Adding the two given angles and subtracting the total from 180 gives an answer of 35 degrees.

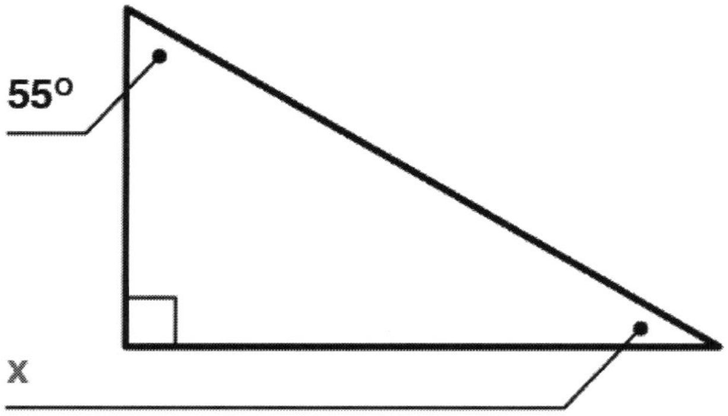

A similar problem can be solved with circles. If the radius is given but the circumference is unknown, the circumference can be calculated based on the formula $C = 2\pi r$. This example can be used in the figure below. The radius can be substituted for r in the formula. Then the circumference can be found as:

$$C = 2\pi \times 8 = 16\pi = 50.24 \text{ cm}$$

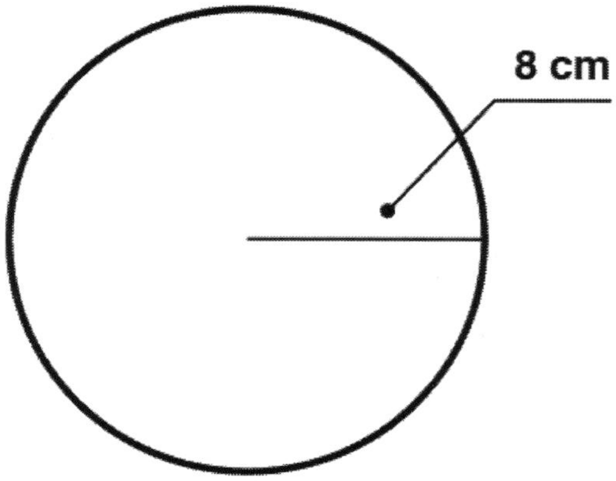

Other figures that may have missing values could be the length of a square, given the area, or the perimeter of a rectangle, given the length and width. All of the missing values can be found by first identifying all the characteristics that are known about the shape, then looking for ways to connect the missing value to the given information.

Using Trigonometric Ratios

Within right triangles, trigonometric ratios can be defined for the acute angle within the triangle. Consider the following right triangle. The side across from the right angle is known as the **hypotenuse,** the acute angle being discussed is labeled $\boldsymbol{\theta}$, the side across from the acute angle is known as the **opposite** side, and the other side is known as the **adjacent** side.

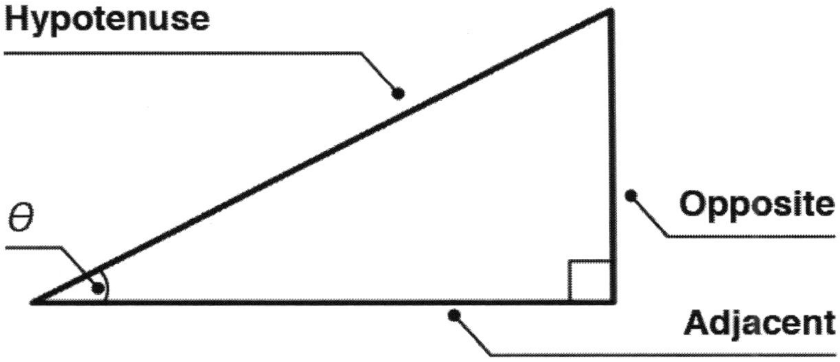

$$\sin\theta = \frac{\text{opposite}}{\text{hypotenuse}} \qquad \csc\theta = \frac{\text{hypotenuse}}{\text{opposite}}$$

$$\cos\theta = \frac{\text{adjacent}}{\text{hypotenuse}} \qquad \sec\theta = \frac{\text{hypotenuse}}{\text{adjacent}}$$

$$\tan\theta = \frac{\text{opposite}}{\text{adjacent}} \qquad \cot\theta = \frac{\text{adjacent}}{\text{opposite}}$$

The six trigonometric ratios are shown above as well. "Sin" is short for sine, "cos" is short for cosine, "tan" is short for tangent, "csc" is short for cosecant, "sec" is short for secant, and "cot" is short for cotangent. A mnemonic device exists that is helpful to remember the ratios. SOHCAHTOA stands for Sine = Opposite/Hypotenuse, Cosine = Adjacent/Hypotenuse, and Tangent = Opposite/Adjacent. The other three trigonometric ratios are reciprocals of sine, cosine, and tangent because:

$$\csc\theta = \frac{1}{\sin\theta}, \sec\theta = \frac{1}{\cos\theta}, \text{ and } \cot\theta = \frac{1}{\tan\theta}$$

The **Pythagorean Theorem** expresses an important relationship between the three sides of a right triangle. It states that the square of the hypotenuse is equal to the sum of the squares of the other two sides. When using the Pythagorean Theorem, the hypotenuse is labeled as side c, the opposite is labeled as side a, and the adjacent side is side b.

The theorem can be seen in the following diagram:

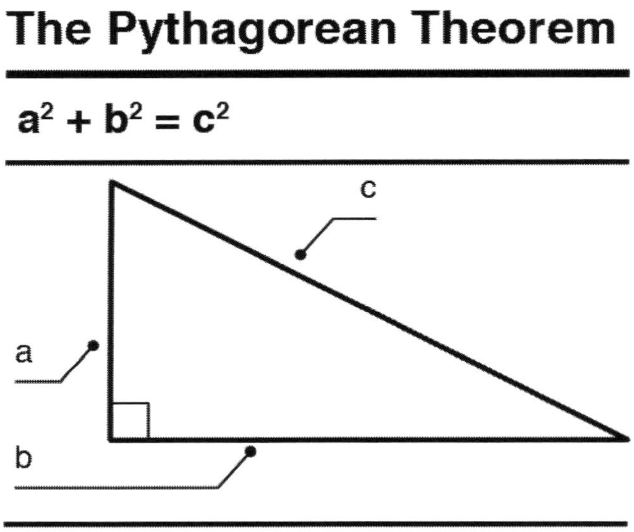

Both the trigonometric ratios and Pythagorean Theorem can be used in problems that involve finding either a missing side or missing angle of a right triangle. Look to see what sides and angles are given and select the correct relationship that will assist in finding the missing value. These relationships can also be used to solve application problems involving right triangles. Often, it is helpful to draw a figure to represent the problem to see what is missing.

Equations of Conic Sections

The intersection of a plane and a double right circular cone is called a **conic section**. There are four basic types of conic sections, a circle, a parabola, a hyperbola, and an ellipse. The equation of a **circle** is given by $(x-h)^2+(y-k)^2=r^2$, where the center of the circle is given by (h,k) and the radius of the circle is r. A parabola that opens up or down has a horizontal axis. The equation of a **parabola with a horizontal axis** is given by $(y-k)^2=4p(x-h)$, where $p \neq 0$ and the vertex is given by (h,k). A parabola that opens to the left or right has a vertical axis. The equation of the **parabola with a vertical axis** is given by $(x-h)^2=4p(y-k)$, where $p \neq 0$ and the vertex is given by (h,k). The equation of an ellipse with a horizontal major axis and center (h,k) is given by:

$$\frac{(x-h)^2}{a^2}+\frac{(y-k)^2}{b^2}=1$$

The distance between center and either focus is c with $c^2=a^2-b^2$, when $a>b>0$. The major axis has length $2a$ and the minor axis has length $2b$. For an ellipse with a vertical major axis and center (h,k), where $a>b>0$, the a and b switch places so the equation is given by:

$$\frac{(x-h)^2}{b^2}+\frac{(y-k)^2}{a^2}=1$$

The major axis still has length $2a$ and the minor axis still has length $2b$, and the distance between center and either focus is $c^2 = a^2 - b^2$, where $a > b > 0$.

A **hyperbola** has an equation similar to the ellipse except that there is a minus in place of the plus sign. A hyperbola with a vertical transverse axis has equation:

$$\frac{(x-h)^2}{a^2} - \frac{(y-k)^2}{b^2} = 1$$

A hyperbola with a horizontal transverse axis has equation:

$$\frac{(y-k)^2}{a^2} - \frac{(x-h)^2}{b^2} = 1$$

For each of these, the center is given by (h, k) and distance between the vertices $2a$.

Statistics & Probability

Describing Center and Spread of Distributions

One way information can be interpreted from tables, charts, and graphs is through statistics. The three most common calculations for a set of data are the mean, median, and mode. These three are called **measures of central tendency**, which are helpful in comparing two or more different sets of data.

The **mean** refers to the average and is found by adding up all values and dividing the total by the number of values. In other words, the mean is equal to the sum of all values divided by the number of data entries. For example, if you bowled a total of 532 points in 4 bowling games, your mean score was $\frac{532}{4} = 133$ points per game. Students can apply the concept of mean to calculate what score they need on a final exam to earn a desired grade in a class.

The **median** is found by lining up values from least to greatest and choosing the middle value. If there is an even number of values, then calculate the mean of the two middle amounts to find the median. For example, the median of the set of dollar amounts \$5, \$6, \$9, \$12, and \$13 is \$9. The median of the set of dollar amounts \$1, \$5, \$6, \$8, \$9, \$10 is \$7, which is the mean of \$6 and \$8.

The **mode** is the value that occurs the most. The mode of the data set {1, 3, 1, 5, 5, 8, 10} actually refers to two numbers: 1 and 5. In this case, the data set is bimodal because it has two modes. A data set can have no mode if no amount is repeated. Another useful statistic is range.

The **range** for a set of data refers to the difference between the highest and lowest value.

In some cases, numbers in a list of data might have weights attached to them. In that case, a **weighted mean** can be calculated. A common application of a weighted mean is GPA. In a semester, each class is assigned a number of credit hours, its weight, and at the end of the semester each student receives a grade. To compute GPA, an A is a 4, a B is a 3, a C is a 2, a D is a 1, and an F is a 0. Consider a student that takes a 4-hour English class, a 3-hour math class, and a 4-hour history class and receives all B's. The weighted mean, GPA, is found by multiplying each grade times its weight, number of credit hours, and dividing by the total number of credit hours.

Therefore, the student's GPA is:

$$\frac{3 \times 4 + 3 \times 3 + 3 \times 4}{11} = \frac{33}{1} = 3.0.$$

The following bar chart shows how many students attend a cycle class on each day of the week. To find the mean attendance for the week, add each day's attendance together:

$$10 + 7 + 6 + 9 + 8 + 14 + 4 = 58$$

Then divide the total by the number of days, $58 \div 7 = 8.3$. The mean attendance for the week was 8.3 people. The median attendance can be found by putting the attendance numbers in order from least to greatest: 4, 6, 7, 8, 9, 10, 14, and choosing the middle number: 8 people. This set of data has no mode because no numbers repeat. The range is 10, which is found by finding the difference between the lowest number, 4, and the highest number, 14.

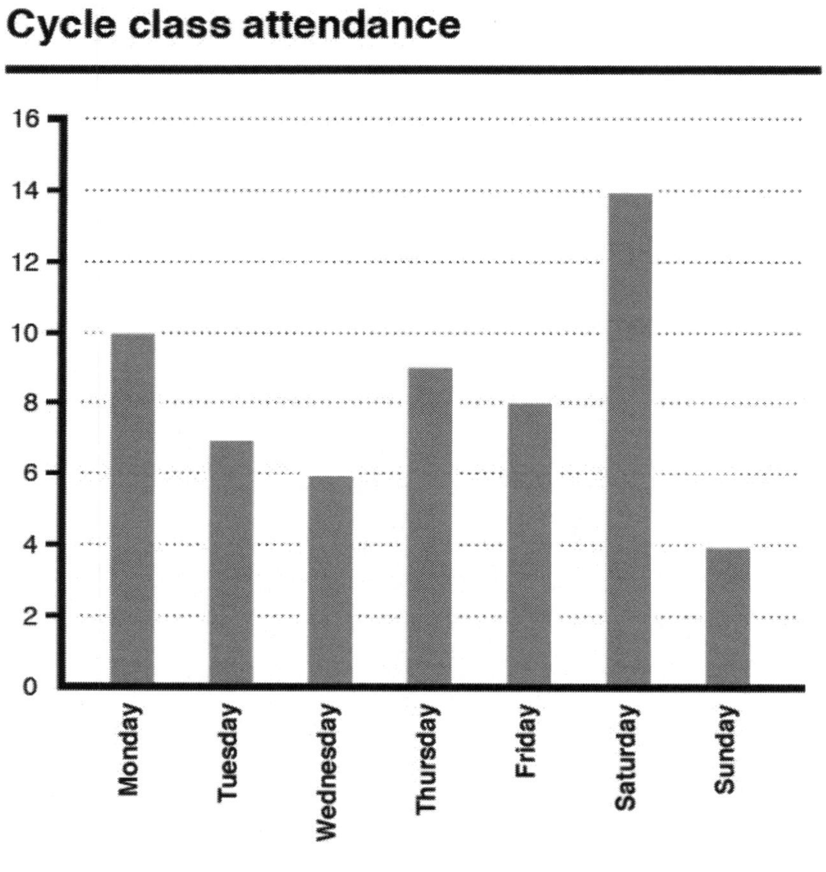

A **histogram** is a bar graph used to group data into "bins" that cover a range on the horizontal, or x-axis. Histograms consist of rectangles whose heights are equal to the frequency of a specific category. The horizontal axis represents the specific categories. Because they cover a range of data, these bins have no gaps between bars, unlike the bar graph above. In a histogram showing the heights of adult golden retrievers, the bottom axis would be groups of heights, and the y-axis would be the number of dogs in each range. Evaluating this histogram would show the height of most golden retrievers as falling within a certain range. It also provides information to find the average height and range for how tall golden retrievers may grow.

The following is a histogram that represents exam grades in a given class. The horizontal axis represents ranges of the number of points scored, and the vertical axis represents the number of students. For example, approximately 33 students scored in the 60 to 70 range.

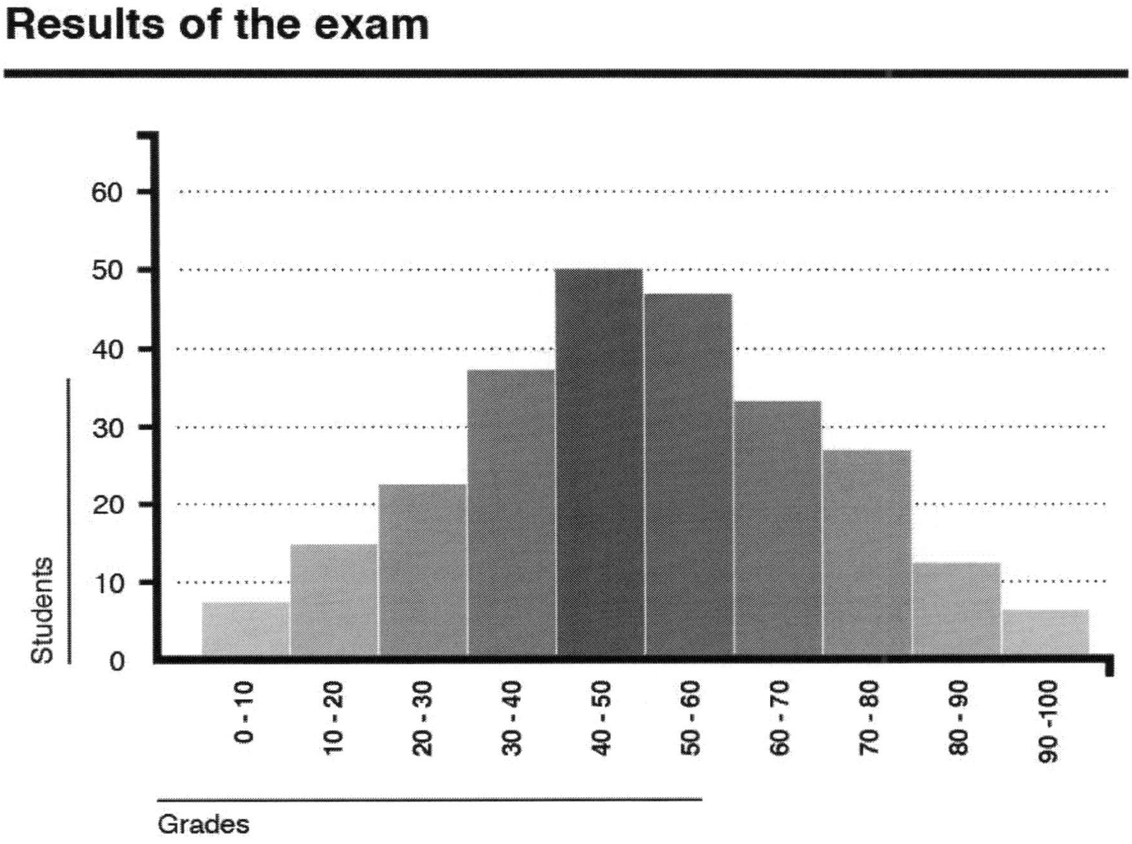

Certain measures of central tendency can be easily visualized with a histogram. If the points scored were shown with individual rectangles, the tallest rectangle would represent the mode. A bimodal set of data would have two peaks of equal height. Histograms can be classified as having data **skewed to the left, skewed to the right,** or **normally distributed**, which is also known as **bell-shaped**. These three classifications can be seen in the following chart:

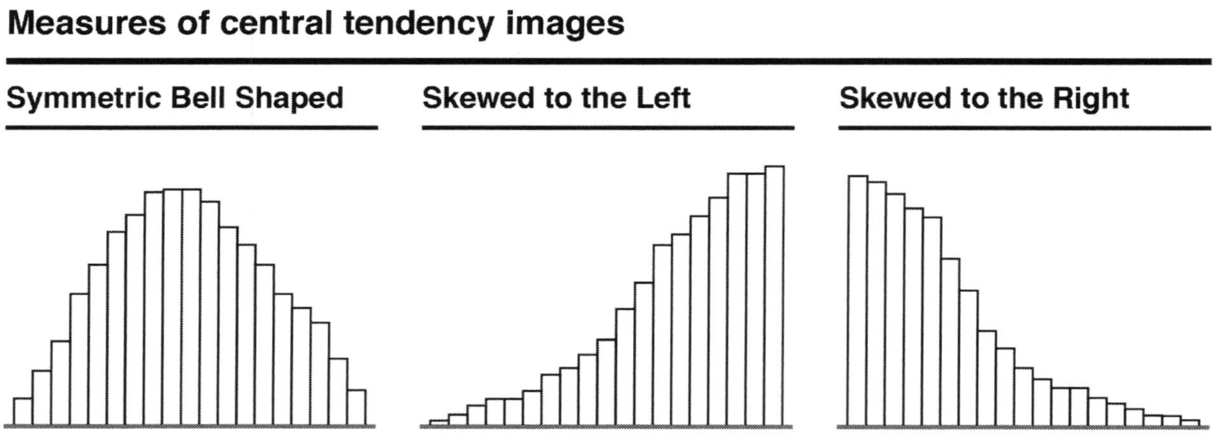

When the data is normal, the mean, median, and mode are very similar because they all represent the most typical value in the data set. In this case, the mean is typically considered the best measure of central tendency because it includes all data points. However, if the data is skewed, the mean becomes less meaningful because it is dragged in the direction of the skew. Therefore, the median becomes the best measure because it is not affected by any outliers.

The measures of central tendency and the range may also be found by evaluating information on a line graph.

The line graph shows the daily high and low temperatures:

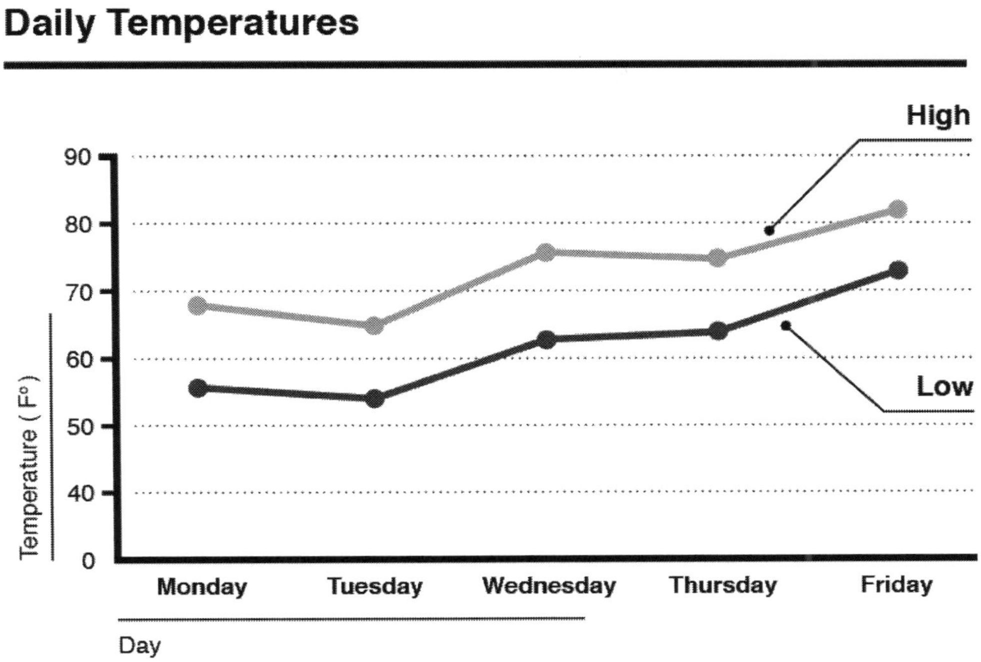

The average high temperature can be found by gathering data from each day on the triangle line. The days' highs are 69, 65, 75, 74, and 81. To find the average, add them together to get 364, then divide by 5 (because there are 5 temperatures). The average high for the five days is 72.8. If 72.8 degrees is found on the graph, it will fall in the middle of all the values. The average low temperature can be found in the same way.

Given a set of data, the **correlation coefficient**, *r*, measures the association between all the data points. If two values are correlated, there is an association between them. However, correlation does not necessarily mean causation, or that one value causes the other. There is a common mistake made that assumes correlation implies causation. Average daily temperature and number of sunbathers are both correlated and have causation. If the temperature increases, that change in weather causes more people to want to catch some rays. However, wearing plus-size clothing and having heart disease are two variables that are correlated but do not have causation. The larger someone is, the more likely he or she is to have heart disease. However, being overweight does not cause someone to have the disease.

The value of the correlation coefficient is between −1 and 1, where −1 represents a perfect negative linear relationship, 0 represents no relationship between the two data sets, and 1 represents a perfect positive linear relationship. A negative linear relationship means that as *x*-values increase, *y*-values decrease. A positive linear relationship means that as *x* values increase, *y* values increase.

The formula for computing the correlation coefficient is:

$$r = \frac{n\sum xy - (\sum x)(\sum y)}{\sqrt{n(\sum x^2) - (\sum x)^2}\sqrt{n(\sum y^2) - (y)^2}}$$

where n is the number of data points. The closer r is to 1 or -1, the stronger the correlation. A correlation can be seen when plotting data. If the graph resembles a straight line, there is a correlation.

Applying and Analyzing Data Collection Methods

Data collection can be done through surveys, experiments, observations, and interviews. A **census** is a type of survey that is done with a whole population. Because it can be difficult to collect data for an entire population, sometimes, a sample survey is used. In this case, one would survey only a fraction of the population and make inferences about the data and generalizations about the larger population from which the sample was drawn. Sample surveys are not as accurate as a census, but this is an easier and less expensive method of collecting data. An **experiment** is used when a researcher wants to explain how one variable causes changes in another variable. For example, if a researcher wanted to know if a particular drug affects weight loss, he or she would choose a treatment group that would take the drug, and another group, the control group, that would not take the drug.

Special care must be taken when choosing these groups to ensure that bias is not a factor. **Bias** occurs when an outside factor influences the outcome of the research. In observational studies, the researcher does not try to influence either variable, but simply observes the behavior of the subjects. Interviews are sometimes used to collect data as well. The researcher will ask questions that focus on her area of interest in order to gain insight from the participants. When gathering data through observation or interviews, it is important that the researcher be well trained so that he or she does not influence the results and so that the study is reliable. A study is reliable if it can be repeated under the same conditions and the same results are received each time.

Understanding and Modeling Relationships in Bivariate Data

Independent and dependent are two types of variables that describe how they relate to each other. The **independent variable** is the variable controlled by the experimenter. It stands alone and is not changed by other parts of the experiment. This variable is normally represented by x and is found on the horizontal, or x-axis, of a graph. The **dependent variable** changes in response to the independent variable. It reacts to, or depends on, the independent variable. This variable is normally represented by y and is found on the vertical, or y-axis of the graph.

The relationship between two variables, x and y, can be seen on a scatterplot.

The following scatterplot shows the relationship between weight and height. The graph shows the weight as x and the height as y. The first dot on the left represents a person who is 45 kg and approximately 150 cm tall. The other dots correspond in the same way. As the dots move to the right and weight increases, height also increases. A line could be drawn through the middle of the dots to move from bottom left to top right. This line would indicate a **positive correlation** between the variables. If the variables had a **negative correlation**, then the dots would move from the top left to the bottom right.

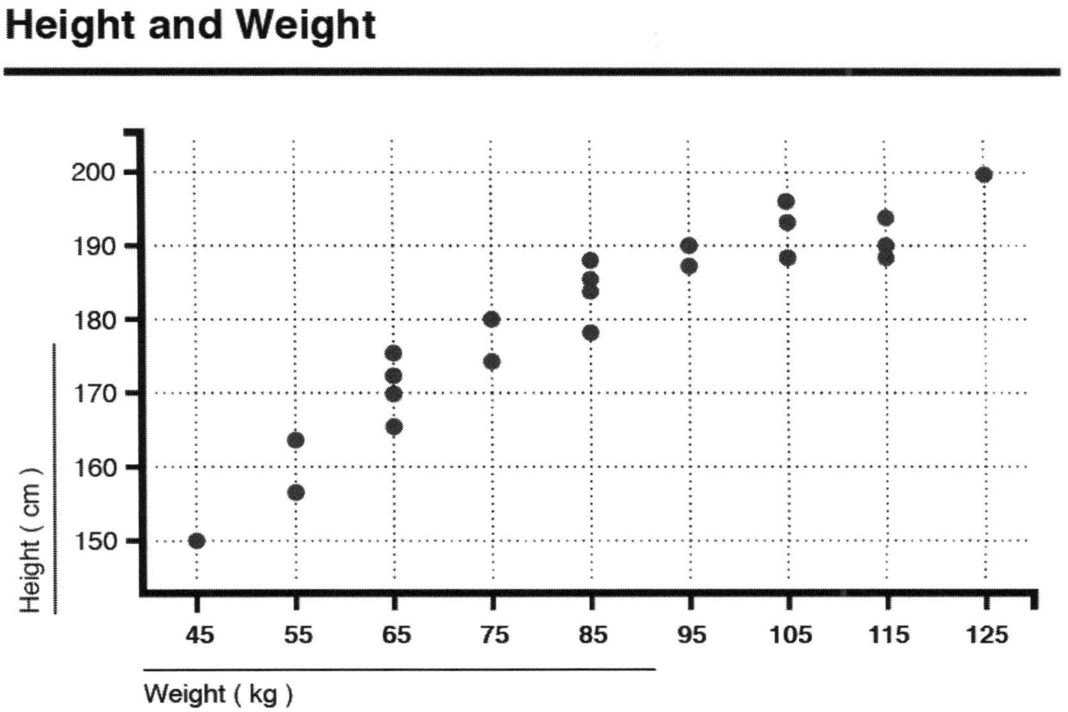

A scatterplot is useful in determining the relationship between two variables, but it is not required. Consider an example where a student scores a different grade on his math test for each week of the month. The independent variable would be the weeks of the month. The dependent variable would be the grades because they change depending on the week. If the grades trended up as the weeks passed, then the relationship between grades and time would be positive. If the grades decreased as the time passed, then the relationship would be negative. (As the number of weeks went up, the grades went down.)

The relationship between two variables can further be described as strong or weak. The relationship between age and height shows a strong positive correlation because children grow taller as they get older. In adulthood, the relationship between age and height becomes weak, and the dots will spread out. People stop growing in adulthood, and their final heights vary depending on factors like genetics and health. The closer the dots on the graph, the stronger the relationship. As they spread apart, the relationship becomes weaker. If they are too spread out to determine a correlation up or down, then the variables are said to have no correlation.

Variables are values that change, so determining the relationship between them requires an evaluation of who or what changes them. If the variable changes because of a result in the experiment, then it's dependent. If the variable changes before the experiment, or is changed by the person controlling the experiment, then it's the independent variable. As they interact, one is manipulated by the other. The manipulator is the independent, and the manipulated is the dependent. Once the independent and dependent variable are determined, they can be evaluated to have a positive, negative, or no correlation.

Calculating Probabilities

Probability describes how likely it is that an event will occur. Probabilities are always a number from zero to 1. If an event has a high likelihood of occurrence, it will have a probability close to 1. If there is only a small chance that an event will occur, the likelihood is close to zero. A fair six-sided die has one of the numbers 1, 2, 3, 4, 5, and 6 on each side. When this die is rolled there is a one in six chance that it will land on 2. This is because there are six possibilities and only one side has a 2 on it. The probability then is $\frac{1}{6}$ or 0.167. The probability of rolling an even number from this die is three in six, which is $\frac{1}{2}$ or 0.5. This is because there are three sides on the die with even numbers (2, 4, 6), and there are six possible sides. The probability of rolling a number less than 10 is 1; since every side of the die has a number less than 6, it would be impossible to roll a number 10 or higher. On the other hand, the probability of rolling a number larger than 20 is zero. There are no numbers greater than 20 on the die, so it is certain that this will not occur, thus the probability is zero.

If a teacher says that the probability of anyone passing her final exam is 0.2, is it highly likely that anyone will pass? No, the probability of anyone passing her exam is low because 0.2 is closer to zero than to 1. If another teacher is proud that the probability of students passing his class is 0.95, how likely is it that a student will pass? It is highly likely that a student will pass because the probability, 0.95, is very close to 1.

Using Two-Way Tables to Summarize Categorical Data and Relative Frequencies, and to Calculate Conditional Probability

A **two-way frequency table** displays categorical data with two variables, and it highlights relationships that exist between those two variables. Such tables are used frequently to summarize survey results, and are also known as **contingency tables**. Each cell shows a count pertaining to that individual variable pairing, known as a **joint frequency**, and the totals of each row and column also are in the table. Consider the following two-way frequency table:

Distribution of the Residents of a Particular Village

	70 or older	69 or younger	Totals
Women	20	40	60
Men	5	35	40
Total	25	75	100

The table above shows the breakdown of ages and sexes of 100 people in a particular village. The end of each row or column displays the number of people represented by the corresponding data, and the total number of people is shown in the bottom right corner. For instance, there were 25 people age 70 or older and 60 women in the data. The 20 in the first cell shows that out of 100 total villagers, 20 were women aged 70 or older. The 5 in the cell below shows that out of 100 total villagers, 5 were men aged 70 or older.

A two-way table can also show relative frequencies by indicating the percentages of people instead of the count. If each frequency is calculated over the entire total of 100, the first cell would be 20% or 0.2. However, the relative frequencies can also be calculated over row or column totals. If row totals were used, the first cell would be:

$$\frac{20}{60} = 0.333 = 33.3\%$$

If column totals were used, the first cell would be:

$$\frac{20}{25} = 0.8 = 80\%$$

Such tables can be used to calculate **conditional probabilities**, which are probabilities that an event occurs, given another event. Consider a randomly selected villager. The probability of selecting a male 70 years old or older is $\frac{5}{100} = 0.05$ because there are 5 males over the age of 70 and 100 total villagers.

Integrating Essential Skills

Rates and Percentages

Ratios and Rates of Change

Recall that a ratio is the comparison of two different quantities. Comparing 2 apples to 3 oranges results in the ratio 2:3, which can be expressed as the fraction $\frac{2}{5}$. Note that order is important when discussing ratios. The number mentioned first is the antecedent, and the number mentioned second is the consequent. Note that the consequent of the ratio and the denominator of the fraction are *not* the same. When there are 2 apples to 3 oranges, there are five fruit total; two fifths of the fruit are apples, while three fifths are oranges. The ratio 2:3 represents a different relationship than the ratio 3:2. Also, it is important to make sure that when discussing ratios that have units attached to them, the two quantities use the same units. For example, to think of 8 feet to 4 yards, it would make sense to convert 4 yards to feet by multiplying by 3. Therefore, the ratio would be 8 feet to 12 feet, which can be expressed as the fraction $\frac{8}{20}$. Also, note that it is proper to refer to ratios in lowest terms. Therefore, the ratio of 8 feet to 4 yards is equivalent to the fraction $\frac{2}{5}$.

Many real-world problems involve ratios. Often, problems with ratios involve proportions, as when two ratios are set equal to find the missing amount. However, some problems involve deciphering single ratios. For example, consider an amusement park that sold 345 tickets last Saturday. If 145 tickets were sold to adults and the rest of the tickets were sold to children, what would the ratio of the number of adult tickets to children's tickets be? A common mistake would be to say the ratio is 145:345. However, 345 is the total number of tickets sold, not the number of children's tickets. There were $345 - 145 = 200$ tickets sold to children. The correct ratio of adult to children's tickets is 145:200. As a fraction, this expression is written as $\frac{145}{345}$, which can be reduced to $\frac{29}{69}$.

While a ratio compares two measurements using the same units, **rates** compare two measurements with different units. Examples of rates would be $200 for 8 hours of work, or 500 miles traveled per 20 gallons. Because the units are different, it is important to always include the units when discussing rates. Key words in rate problems include for, per, on, from, and in. Just as with ratios, it is important to write rates in lowest terms. A common rate in real-life situations is cost per unit, which describes how much one item/unit costs. When evaluating the cost of an item that comes in several sizes, the cost per unit rate can help buyers determine the best deal. For example, if 2 quarts of soup was sold for $3.50 and 3 quarts was sold for $4.60, to determine the best buy, the cost per quart should be found. $\frac{\$3.50}{2\text{ qt}} = \$1.75$ per quart, and $\frac{\$4.60}{3\text{ qt}} = \$1.53$ per quart. Therefore, the better deal would be the 3-quart option.

Rate of change problems involve calculating a quantity per some unit of measurement. Usually the unit of measurement is time. For example, meters per second is a common rate of change. To calculate this measurement, find the distance traveled in meters and divide by total time traveled. The result is the average speed over the entire time interval. Another common rate of change used in the real world is miles per hour. Consider the following problem that involves calculating an average rate of change in temperature. Last Saturday, the temperature at 1:00 a.m. was 34 degrees Fahrenheit, and at noon, the temperature had increased to 75 degrees Fahrenheit. What was the average rate of change over that time interval? The average rate of change is calculated by finding the total change in temperature and dividing by the total hours elapsed. Therefore, the rate of change was equal to:

$$\frac{75-34}{12-1} = \frac{41}{11} \text{ degrees per hour}$$

This quantity, rounded to two decimal places, is equal to 3.72 degrees per hour.

A common rate of change that appears in algebra is the slope calculation. Given a linear equation in one variable, $y = mx + b$, the **slope**, m, is equal to $\frac{rise}{run}$ or $\frac{change\ in\ y}{change\ in\ x}$. In other words, slope is equivalent to the ratio of the vertical and horizontal changes between any two points on a line. The vertical change is known as the **rise**, and the horizontal change is known as the **run**. Given any two points on a line (x_1, y_1) and (x_2, y_2), slope can be calculated with the formula:

$$m = \frac{y_2 - y_1}{x_2 - x_1} = \frac{\Delta y}{\Delta x}$$

Common real-world applications of slope include determining how steep a staircase should be, calculating how steep a road is, and determining how to build a wheelchair ramp.

Many times, problems involving rates and ratios involve proportions. A proportion states that two ratios (or rates) are equal. The property of cross products can be used to determine if a proportion is true, meaning both ratios are equivalent. If $\frac{a}{b} = \frac{c}{d}$, then to clear the fractions, multiply both sides by the least common denominator, bd. This results in $ad = bc$, which is equal to the result of multiplying along both diagonals. For example, $\frac{4}{40} = \frac{1}{10}$ grants the cross product $4 \times 10 = 40 \times 1$, which is equivalent to $40 = 40$ and shows that this proportion is true. Cross products are used when proportions are involved in real-world problems. Consider the following: If 3 pounds of fertilizer will cover 75 square feet of grass, how many pounds are needed for 375 square feet? To solve this problem, set up a proportion using two ratios. Let x equal the unknown quantity, pounds needed for 375 feet.

Setting the two ratios equal to one another yields the equation:

$$\frac{3}{75} = \frac{x}{375}$$

Cross-multiplication gives:

$$3 \times 375 = 75x$$

Therefore, $1{,}125 = 75x$. Divide both sides by 75 to get $x = 15$. Therefore, 15 pounds of fertilizer are needed to cover 375 square feet of grass.

Another application of proportions involves similar triangles. If two triangles have corresponding angles with the same measurements and corresponding sides with proportional measurements, the triangles are said to be similar. If two angles are the same, the third pair of angles are equal as well because the sum of all angles in a triangle is equal to 180 degrees. Each pair of equivalent angles are known as **corresponding angles. Corresponding sides** face the corresponding angles, and it is true that corresponding sides are in proportion.

For example, consider the following set of similar triangles:

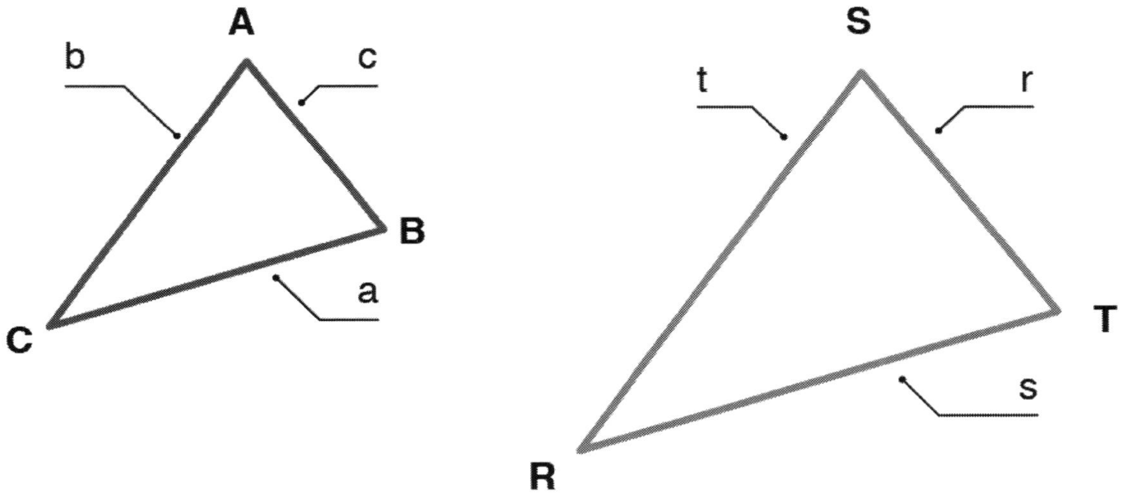

Angles A and S have the same measurement, angles C and R have the same measurement, and angles B and T have the same measurement. Therefore, the following proportion can be set up from the sides:

$$\frac{c}{r} = \frac{a}{s} = \frac{b}{t}$$

This proportion can be helpful in finding missing lengths in pairs of similar triangles. For example, if the following triangles are similar, a proportion can be used to find the missing side lengths, a and b.

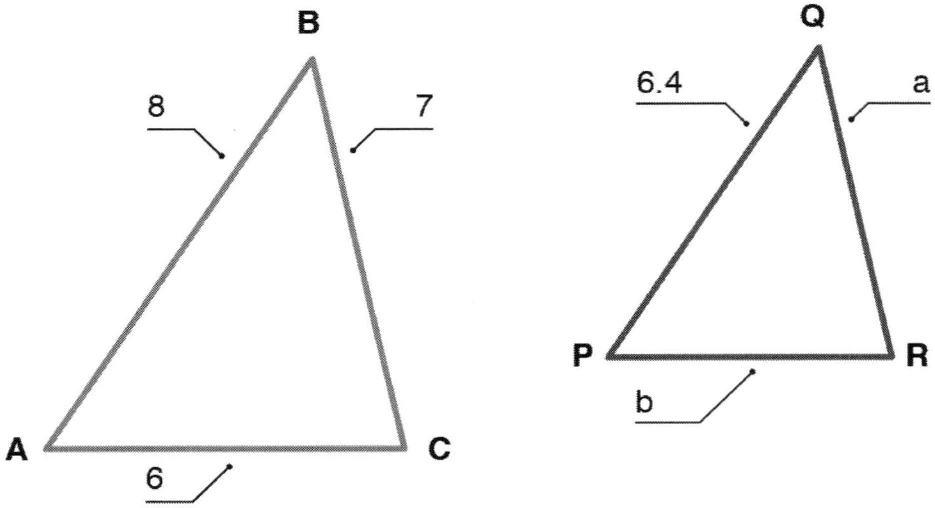

The proportions $\frac{8}{6.4} = \frac{6}{b}$ and $\frac{8}{6.4} = \frac{7}{a}$ can both be cross multiplied and solved to obtain a = 5.6 and b = 4.8.

A real-life situation that uses similar triangles involves measuring shadows to find heights of unknown objects. Consider the following problem: A building casts a shadow that is 120 feet long, and at the same time, another building that is 80 feet high casts a shadow that is 60 feet long. How tall is the first building? Each building, together with the sun rays and shadows casted on the ground, forms a triangle. They are similar because each building forms a right angle with the ground, and the sun rays form equivalent angles. Therefore, these two pairs of angles are both equal. Because all angles in a triangle add up to 180 degrees, the third angles are equal as well. Both shadows form corresponding sides of the triangle, the buildings form corresponding sides, and the sun rays form corresponding sides. Therefore, the triangles are similar, and the following proportion can be used to find the missing building length:

$$\frac{120}{x} = \frac{60}{80}$$

Cross multiply to obtain the equation $9600 = 60x$. Then, divide both sides by 60 to obtain $x = 160$. This means that the first building is 160 feet high.

Percentages

Percentages are defined as parts per one hundred. To convert a decimal to a percentage, move the decimal point two units to the right and place the percent sign after the number. Percentages appear in many scenarios in the real world. It is important to make sure the statement containing the percentage is translated to a correct mathematical expression. Be aware that it is extremely common to make a mistake when working with percentages within word problems.

An example of a word problem containing a percentage is the following: 35% of people speed when driving to work. In a group of 5,600 commuters, how many would be expected to speed on the way to their place of employment? The answer to this problem is found by finding 35% of 5,600. First, change the percentage to the decimal 0.35. Then compute the product: $0.35 \times 5{,}600 = 1{,}960$. Therefore, it would be expected that 1,960 of those commuters would speed on their way to work based on the data given. In this situation, the word "of" signals to use multiplication to find the answer. Another way percentages are used is in the following problem: Teachers work 8 months out of the year. What percent of the year do they work? To answer this problem, find what percent of 12 the number 8 is, because there are 12 months in a year. Therefore, divide 8 by 12, and convert that number to a percentage:

$$\frac{8}{12} = \frac{2}{3} = 0.66\overline{6}$$

The percentage rounded to the nearest tenth place tells us that teachers work 66.7% of the year. Percentage problems can also find missing quantities like in the following question: 60% of what number is 75? To find the missing quantity, turn the question into an equation. Let x be equal to the missing quantity. Therefore, $0.60x = 75$. Divide each side by 0.60 to obtain 125. Therefore, 60% of 125 is equal to 75.

Sales tax is an important application relating to percentages because tax rates are usually given as percentages. For example, a city might have an 8% sales tax rate. Therefore, when an item is purchased with that tax rate, the real cost to the customer is 1.08 times the price in the store. For example, a $25 pair of jeans costs the customer:

$$\$25 \times 1.08 = \$27$$

If the sales tax rate is unknown, it can be determined after an item is purchased. If a customer visits a store and purchases an item for $21.44, but the price in the store was $19, they can find the tax rate by first subtracting $\$21.44 - \$19$ to obtain $2.44, the sales tax amount. The sales tax is a percentage of the in-store price. Therefore, the tax rate is $\frac{2.44}{19} = 0.128$, which has been rounded to the nearest thousandths place. In this scenario, the actual sales tax rate given as a percentage is 12.8%.

Solving Unit Rate Problems

A **unit rate** is a rate with a denominator of one. It is a comparison of two values with different units where one value is equal to one. Examples of unit rates include 60 miles per hour and 200 words per minute. Problems involving unit rates may require some work to find the unit rate. For example, if Mary travels 360 miles in 5 hours, what is her speed, expressed as a unit rate? The rate can be expressed as the following fraction:

$$\frac{360\ miles}{5\ hours}$$

The denominator can be changed to one by dividing by five. The numerator will also need to be divided by five to follow the rules of equality. This division turns the fraction into $\frac{72\ miles}{1\ hour}$, which can now be labeled as a unit rate because one unit has a value of one. Another type question involves the use of unit rates to solve problems. For example, if Trey needs to read 300 pages and his average speed is 75 pages per hour, will he be able to finish the reading in 5 hours? The unit rate is 75 pages per hour, so the total of 300 pages can be divided by 75 to find the time. After the division, the time it takes to read is four hours. The answer to the question is yes, Trey will finish the reading within 5 hours.

Proportional Relationships

Fractions appear in everyday situations, and in many scenarios, they appear in the real-world as ratios and in proportions. A **ratio** is formed when two different quantities are compared. For example, in a group of

50 people, if there are 33 females and 17 males, the ratio of females to males is 33 to 17. This expression can be written in the fraction form as $\frac{33}{50}$, where the denominator is the sum of females and males, or by using the ratio symbol, 33:17. The order of the number matters when forming ratios. In the same setting, the ratio of males to females is 17 to 33, which is equivalent to $\frac{17}{50}$ or 17:33. A **proportion** is an equation involving two ratios. The equation $\frac{a}{b} = \frac{c}{d}$, or $a{:}\,b = c{:}\,d$ is a proportion, for real numbers a, b, c, and d. Usually, in one ratio, one of the quantities is unknown, and cross-multiplication is used to solve for the unknown. Consider,

$$\frac{1}{4} = \frac{x}{5}$$

To solve for x, cross-multiply to obtain $5 = 4x$. Divide each side by 4 to obtain the solution $x = \frac{5}{4}$. It is also true that percentages are ratios in which the second term is 100 minus the first term. For example, 65% is 65:35 or $\frac{65}{100}$. Therefore, when working with percentages, one is also working with ratios.

Real-world problems frequently involve proportions. For example, consider the following problem: If 2 out of 50 pizzas are usually delivered late from a local Italian restaurant, how many would be late out of 235 orders? The following proportion would be solved with x as the unknown quantity of late pizzas:

$$\frac{2}{50} = \frac{x}{235}$$

Cross multiplying results in $470 = 50x$. Divide both sides by 50 to obtain $x = \frac{470}{50}$, which in lowest terms is equal to $\frac{47}{5}$. In decimal form, this improper fraction is equal to 9.4. Because it does not make sense to answer this question with decimals (portions of pizzas do not get delivered) the answer must be rounded. Traditional rounding rules would say that 9 pizzas would be expected to be delivered late. However, to be safe, rounding up to 10 pizzas out of 235 would probably make more sense.

Area, Surface Area, and Volume

Perimeter and **area** are geometric quantities that describe objects' measurements. **Perimeter** is the distance around an object. The perimeter of an object can be found by adding the lengths of all sides. Perimeter may be used in problems dealing with lengths around objects such as fences or borders. It may also be used in finding missing lengths or working backwards. If the perimeter is given, but a length is missing, use subtraction to find the missing length.

Given a square with side length s, the formula for perimeter is $P = 4s$. Given a rectangle with length l and width w, the formula for perimeter is:

$$P = 2l + 2w$$

The perimeter of a triangle is found by adding the three side lengths, and the perimeter of a trapezoid is found by adding the four side lengths. The units for perimeter are always the original units of length, such as meters, inches, miles, etc. When discussing a circle, the distance around the object is referred to as its **circumference,** not perimeter. The formula for the circumference of a circle is $C = 2\pi r$, where r represents the radius of the circle. This formula can also be written as $C = d\pi$, where d represents the diameter of the circle.

Area is the two-dimensional space covered by an object. These problems may include the area of a rectangle, a yard, or a wall to be painted. Finding the area may require a simple formula or multiple formulas used together. The units for area are square units, such as square meters, square inches, and square miles. Given a square with side length s, the formula for its area is $A = s^2$.

The table below shows some other common shapes and their area formulas:

Shape	Formula	Graphic
Rectangle	$Area = length \times width$	
Triangle	$Area = \frac{1}{2} \times base \times height$	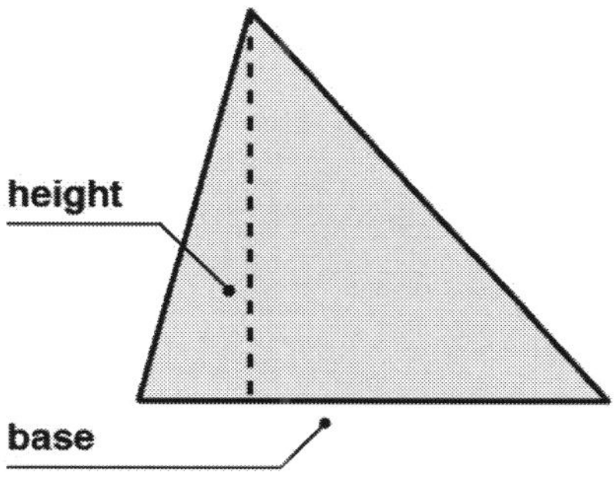
Circle	$Area = \pi \times radius^2$	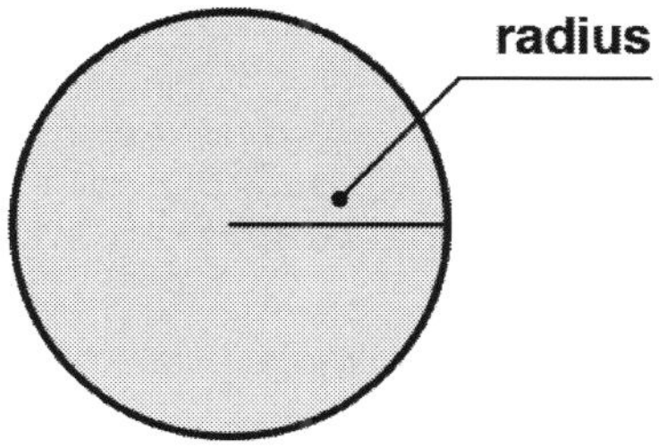

To find the area of the shapes above, use the given dimensions of the shape in the formula.

The following formula, not as widely used as those shown above, but very important, is the area of a trapezoid:

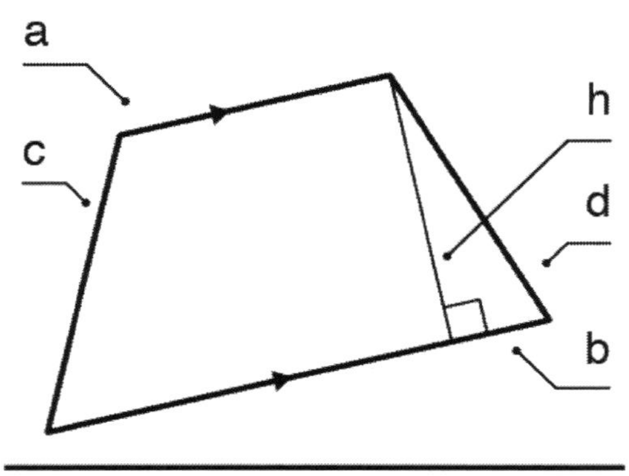

$$A = \frac{1}{2}(a + b)h$$

Complex shapes might require more than one formula. To find the area of the figure below, break the figure into two shapes.

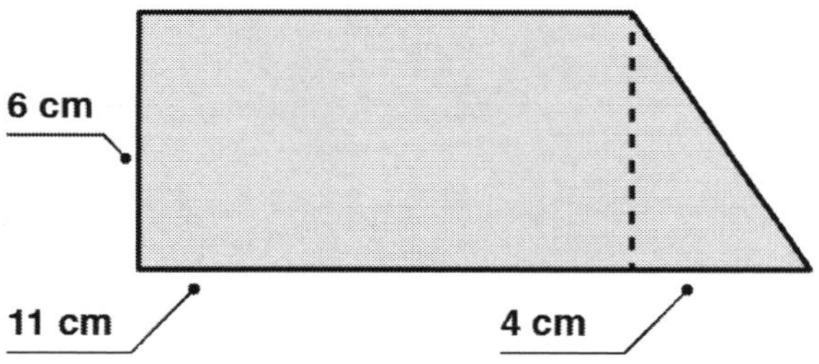

The rectangle has dimensions 6 cm by 11 cm. The triangle has a base of 4 cm and a height of 6 cm. Plug the dimensions into the rectangle formula:

$$A = 6 \times 11$$

Multiplication yields an area of 66 cm^2. The triangle's area can be found using the formula:

$$A = \frac{1}{2} \times 4 \times 6$$

Multiplication yields an area of 12 cm^2.

Add the two areas to find the total area area of the figure, which is 78 cm^2.

Instead of combining areas, some problems may require subtracting them, or finding the difference. To find the area of the shaded region in the figure below, determine the area of the whole figure. Then subtract the area of the circle from the whole.

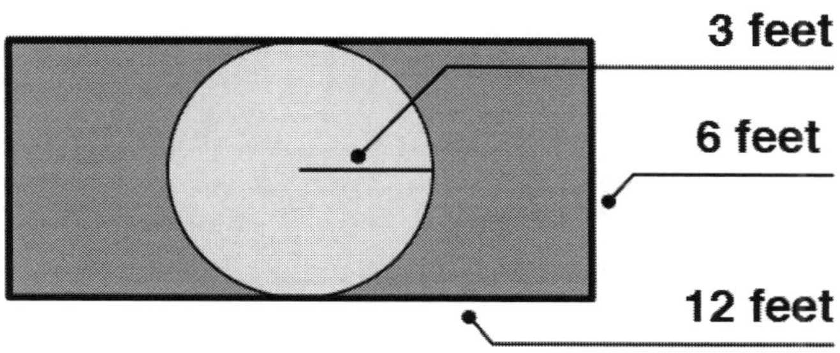

The following formula shows the area of the outside rectangle:

$$A = 12 \times 6 = 72 \text{ ft}^2$$

The area of the inside circle can be found by the following formula:

$$A = \pi(3)^2 = 9\pi = 28.3 \text{ ft}^2$$

As the shaded area is outside the circle, the area for the circle can be subtracted from the area of the rectangle to yield an area of 43.7 ft^2.

Geometric figures may be shown as pictures or described in words. If a rectangular playing field with dimensions 95 meters long by 50 meters wide is measured for perimeter, the distance around the field must be found. The perimeter includes two lengths and two widths to measure the entire outside of the field. This quantity can be calculated using the following equation:

$$P = 2(95) + 2(50) = 290\ m$$

The distance around the field is 290 meters.

Perimeter and area are two-dimensional descriptions; volume is three-dimensional. Volume describes the amount of space that an object occupies, but it differs from area because it has three dimensions instead of two. The units for volume are cubic units, such as cubic meters, cubic inches, and cubic millimeters. Volume can be found by using formulas for common objects such as cylinders and boxes.

The following chart shows a diagram and formula for the volume of two objects.

Shape	Formula	Diagram
Rectangular Prism (box)	$V = length \times width \times height$	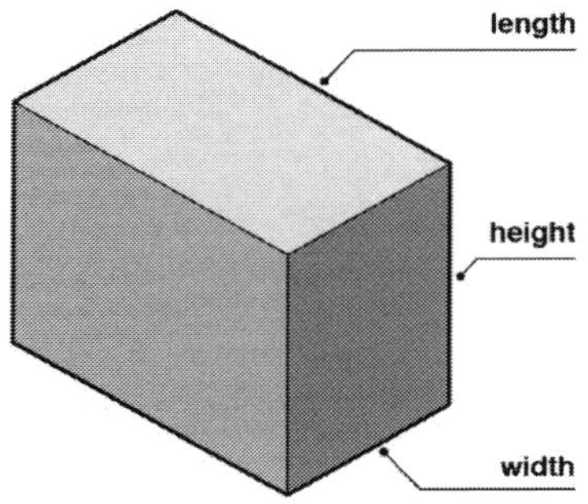
Cylinder	$V = \pi \times radius^2 \times height$	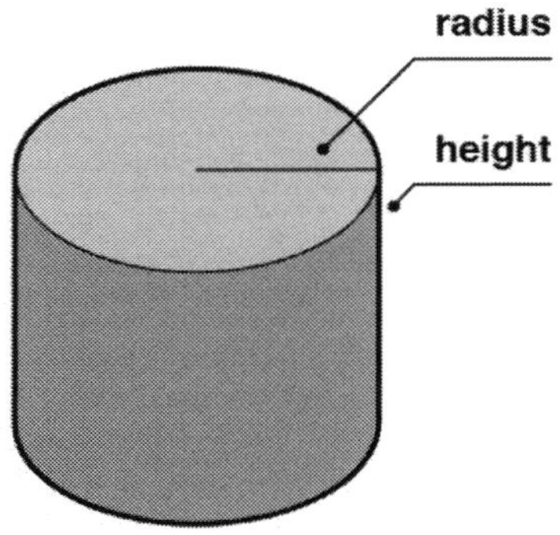

Volume formulas of these two objects are derived by finding the area of the bottom two-dimensional shape, such as the circle or rectangle, and then multiplying times the height of the three-dimensional shape. Other volume formulas include the volume of a cube with side length s: $V = s^3$; the volume of a sphere with radius r: $V = \frac{4}{3}\pi r^3$; and the volume of a cone with radius r and height h:

$$V = \frac{1}{3}\pi r^2 h$$

If a soda can has a height of 5 inches and a radius on the top of 1.5 inches, the volume can be found using one of the given formulas. A soda can is a cylinder. Knowing the given dimensions, the formula can be completed as follows:

$$V = \pi(radius)^2 \times height$$

$$\pi(1.5 \text{ in})^2 \times 5 \text{ in} = 35.325 \text{ in}^3$$

Notice that the units for volume are inches cubed because it refers to the number of cubic inches required to fill the can.

With any geometric calculations, it's important to determine what dimensions are given and what quantities the problem is asking for. If a connection can be made between them, the answer can be found.

Other geometric quantities can include angles inside a triangle. The sum of the measures of any triangle's three angles is 180 degrees. Therefore, if only two angles are known, the third can be found by subtracting the sum of the two known quantities from 180. Two angles that add up to 90 degrees are known as complementary angles. For example, angles measuring 72 and 18 degrees are complementary. Finally, two angles that add up to 180 degrees are known as supplementary angles. To find the supplement of an angle, subtract the given angle from 180 degrees. For example, the supplement of an angle that is 50 degrees is:

$$180 - 50 = 130 \text{ degrees}$$

These terms involving angles can be seen in many types of word problems. For example, consider the following problem: The measure of an angle is 60 degrees less than two times the measure of its complement. What is the angle's measure? To solve this, let x be the unknown angle. Therefore, its complement is $90 - x$. The problem gives that:

$$x = 2(90 - x) - 60$$

To solve for x, distribute the 2, and collect like terms. This process results in:

$$x = 120 - 2x$$

Then, use the addition property to add $2x$ to both sides to obtain:

$$3x = 120$$

Finally, use the multiplication properties of equality to divide both sides by 3 to get:

$$x = 40$$

Therefore, the angle measures 40 degrees. Also, its complement measures 50 degrees.

Determining Surface Area Measurements

As mentioned previously, **surface area** is defined as the area of the surface of a figure. A pyramid has a surface made up of four triangles and one square. To calculate the surface area of a pyramid, the areas of each individual shape are calculated. Then the areas are added together. This method of decomposing the shape into two-dimensional figures to find area, then adding the areas, can be used to find surface area for any figure. Once these measurements are found, the area is described with square units. The surface area of a cube is found by multiplying the area of one side by six since there are six equivalent sides. For example, if the side length is 4 inches, the surface area is found using:

$$SA = 6 \times (4 \times 4) = 96 \text{ square inches}$$

Average and Median

A data set can be described by calculating the mean, median, and mode. These values, called **measures of center,** allow the data to be described with a single value that is representative of the data set.

Again, the most common measure of center is the **mean,** also referred to as the **average.**

To calculate the mean:

- Add all data values together
- Divide by the sample size (the number of data points in the set)

The **median** is middle data value, so that half of the data lies below this value and half lies above it.

To calculate the median,

- Order the data from least to greatest
- The point in the middle of the set is the median
    - In the event that there is an even number of data points, add the two middle points and divide by 2

The **mode** is the data value that occurs most often.

To calculate the mode,

- Order the data from least to greatest
- Find the value that occurs most often

Example: Amelia is a leading scorer on the school's basketball team. The following data set represents the number of points that Amelia has scored in each game this season. Use the mean, median, and mode to describe the data.

16, 12, 26, 14, 28, 14, 12, 15, 25

Solution:

Mean:

$$16 + 12 + 26 + 14 + 28 + 14 + 12 + 15 + 25 = 162$$

$$162 \div 9 = 18$$

Amelia averages 18 points per game.

Median:

12, 12, 14, 14, **15**, 16, 25, 26, 28

Amelia's median score is 15.

Mode:

12, 12, 14, 14, 15, 16, 25, 26, 28

The numbers 12 and 14 each occur twice in the data set, so this set has 2 modes: 12 and 14.

The **range** is the difference between the largest and smallest values in the set. In the example above, the range is:

$$28 - 12 = 16$$

Expressing Numbers in Different Ways

Converting Non-Negative Fractions, Decimals, and Percentages

Within the number system, different forms of numbers can be used. It is important to be able to recognize each type, as well as work with, and convert between, the given forms. The **real number system** comprises natural numbers, whole numbers, integers, rational numbers, and irrational numbers. Natural numbers, whole numbers, integers, and irrational numbers typically are not represented as fractions, decimals, or percentages. Rational numbers, however, can be represented as any of these three forms. A **rational number** is a number that can be written in the form $\frac{a}{b}$, where a and b are integers, and b is not equal to zero. In other words, rational numbers can be written in a fraction form. The value a is the **numerator**, and b is the **denominator.**

If the numerator is equal to zero, the entire fraction is equal to zero. Non-negative fractions can be less than 1, equal to 1, or greater than 1. Fractions are less than 1 if the numerator is smaller (less than) than the denominator. For example, $\frac{3}{4}$ is less than 1. A fraction is equal to 1 if the numerator is equal to the denominator. For instance, $\frac{4}{4}$ is equal to 1. Finally, a fraction is greater than 1 if the numerator is greater than the denominator: the fraction $\frac{11}{4}$ is greater than 1.

When the numerator is greater than the denominator, the fraction is called an **improper fraction**. An improper fraction can be converted to a **mixed number,** a combination of both a whole number and a fraction. To convert an improper fraction to a mixed number, divide the numerator by the denominator. Write down the whole number portion, and then write any remainder over the original denominator. For example, $\frac{11}{4}$ is equivalent to $2\frac{3}{4}$. Conversely, a mixed number can be converted to an improper fraction by multiplying the denominator by the whole number and adding that result to the numerator.

Fractions can be converted to decimals. With a calculator, a fraction is converted to a decimal by dividing the numerator by the denominator. For example:

$$\frac{2}{5} = 2 \div 5 = 0.4$$

Sometimes, rounding might be necessary. Consider:

$$\frac{2}{7} = 2 \div 7 = 0.28571429$$

This decimal could be rounded for ease of use, and if it needed to be rounded to the nearest thousandth, the result would be 0.286. If a calculator is not available, a fraction can be converted to a decimal manually. First, find a number that, when multiplied by the denominator, has a value equal to 10, 100, 1,000, etc. Then, multiply both the numerator and denominator times that number. The decimal form of the fraction is equal to the new numerator with a decimal point placed as many place values to the left as there are zeros in the denominator. For example, to convert $\frac{3}{5}$ to a decimal, multiply both the numerator and denominator times 2, which results in $\frac{6}{10}$. The decimal is equal to 0.6 because there is one zero in the denominator, and so the decimal place in the numerator is moved one unit to the left.

In the case where rounding would be necessary while working without a calculator, an approximation must be found. A number close to 10, 100, 1,000, etc. can be used. For example, to convert $\frac{1}{3}$ to a decimal, the numerator and denominator can be multiplied by 33 to turn the denominator into approximately 100, which makes for an easier conversion to the equivalent decimal. This process results in $\frac{33}{99}$ and an approximate decimal of 0.33. Once in decimal form, the number can be converted to a percentage. Multiply the decimal by 100 and then place a percent sign after the number. For example, 0.614 is equal to 61.4%. In other words, move the decimal place two units to the right and add the percentage symbol.

Composing and Decomposing Multidigit Numbers

Composing and decomposing numbers reveals the place value held by each number 0 through 9 in each position. For example, the number 17 is read as "seventeen." It can be decomposed into the numbers 10 and 7. It can be described as 1 group of ten and 7 ones. The one in the tens place represents one set of ten. The seven in the ones place represents seven sets of one. Added together, they make a total of seventeen. The number 48 can be written in words as "forty-eight." It can be decomposed into the numbers 40 and 8, where there are 4 groups of ten and 8 groups of one. The number 296 can be decomposed into 2 groups of one hundred, 9 groups of ten, and 6 groups of one. There are two hundreds, nine tens, and six ones. Decomposing and composing numbers lays the foundation for visually picturing the number and its place value, and adding and subtracting multiple numbers with ease.

Modeling

Producing, Interpreting, Understanding, Evaluating, and Improving Models

Interpreting Relevant Information from Tables, Charts, and Graphs

They all organize, categorize, and compare data, and they come in different shapes and sizes. Each type has its own way of showing information, whether through a column, shape, or picture. To answer a question relating to a table, chart, or graph, some steps should be followed. First, the problem should be read thoroughly to determine what is being asked to determine what quantity is unknown. Then, the title of the table, chart, or graph should be read. The title should clarify what data is actually being summarized in the table. Next, look at the key and labels for both the horizontal and vertical axes, if they are given. These items will provide information about how the data is organized. Finally, look to see if there is any more labeling inside the table. Taking the time to get a good idea of what the table is summarizing will be helpful as it is used to interpret information.

Tables are a good way of showing a lot of information in a small space. The information in a table is organized in columns and rows. For example, a table may be used to show the number of votes each candidate received in an election. By interpreting the table, one may observe which candidate won the election and which candidates came in second and third. In using a bar chart to display monthly rainfall amounts in different countries, rainfall can be compared between counties at different times of the year. Graphs are also a useful way to show change in variables over time, as in a line graph, or percentages of a whole, as in a pie graph.

The table below relates the number of items to the total cost. The table shows that one item costs $5. By looking at the table further, five items cost $25, ten items cost $50, and fifty items cost $250. This cost can be extended for any number of items. Since one item costs $5, then two items would cost $10. Though this information is not in the table, the given price can be used to calculate unknown information.

Number of Items	1	5	10	50
Cost ($)	5	25	50	250

A bar graph is a graph that summarizes data using bars of different heights. It is useful when comparing two or more items or when seeing how a quantity changes over time. It has both a horizontal and vertical axis. To interpret bar graphs, recognize what each bar represents and connect that to the two variables. The bar graph below shows the scores for six people during three different games. The different colors of the bars distinguish between the three games, and the height of the bar indicates their score for that game. William scored 25 on game 3, and Abigail scored 38 on game 3. By comparing the bars, it is obvious that Williams scored lower than Abigail.

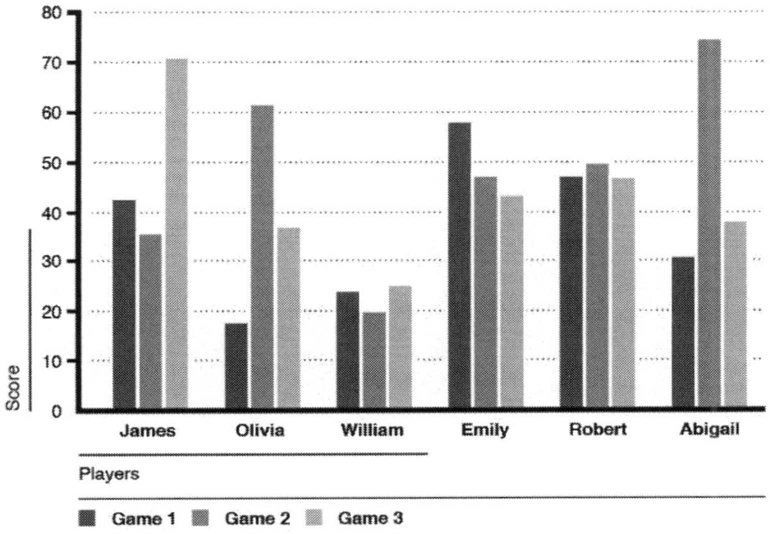

A line graph is a way to compare two variables that are plotted on opposite axes of a graph. The line indicates a continuous change as it rises or falls. The line's rate of change is known as its slope. The horizontal axis often represents a variable of time. Audiences can quickly see if an amount has grown or decreased over time. The bottom of the graph, or the x-axis, shows the units for time, such as days, hours, months, etc. If there are multiple lines, a comparison can be made between what the two lines represent. For example, as shown previously, the following line graph shows the change in temperature over five days.

The top line represents the high, and the bottom line represents the low for each day. Looking at the top line alone, the high decreases for a day, then increases on Wednesday. Then it decreases on Thursday and increases again on Friday. The low temperatures have a similar trend, shown in the bottom line. The range in temperatures each day can also be calculated by finding the difference between the top line and bottom line on a particular day. On Wednesday, the range was 14 degrees, from 62 to 76° F.

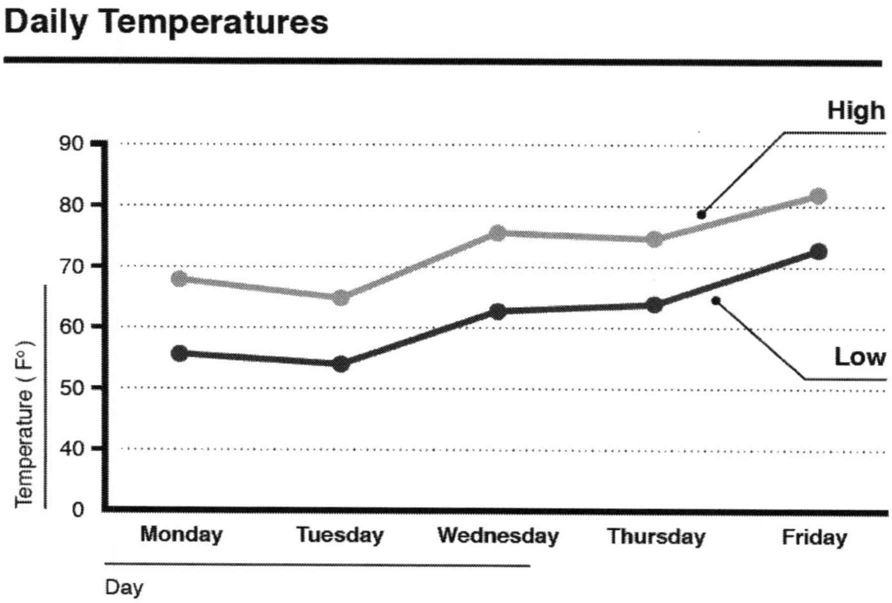

Pie charts are used to show percentages of a whole, as each category is given a piece of the pie, and together all the pieces make up a whole. They are a circular representation of data which are used to highlight numerical proportion. It is true that the arc length of each pie slice is proportional to the amount it individually represents. When a pie chart is shown, an audience can quickly make comparisons by comparing the sizes of the pieces of the pie. They can be useful for comparison between different categories. The following pie chart is a simple example of three different categories shown in comparison to each other.

Light gray represents cats, dark gray represents dogs, and the medium shade of gray represents other pets. These three equal pieces each represent just more than 33 percent, or $\frac{1}{3}$ of the whole. Values 1 and 2 may be combined to represent $\frac{2}{3}$ of the whole. In an example where the total pie represents 75,000 animals, then cats would be equal to $\frac{1}{3}$ of the total, or 25,000. Dogs would equal 25,000 and other pets would also equal 25,000.

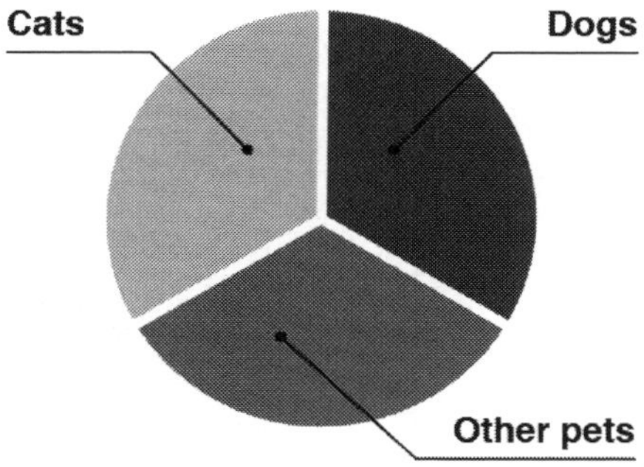

Since circles have 360 degrees, they are used to create pie charts. Because each piece of the pie is a percentage of a whole, that percentage is multiplied times 360 to get the number of degrees each piece represents. In the example above, each piece is $\frac{1}{3}$ of the whole, so each piece is equivalent to 120 degrees. Together, all three pieces add up to 360 degrees.

Stacked bar graphs, also used fairly frequently, are used when comparing multiple variables at one time. They combine some elements of both pie charts and bar graphs, using the organization of bar graphs and the proportionality aspect of pie charts. The following is an example of a stacked bar graph that represents the number of students in a band playing drums, flute, trombone, and clarinet. Each bar graph is broken up further into girls and boys.

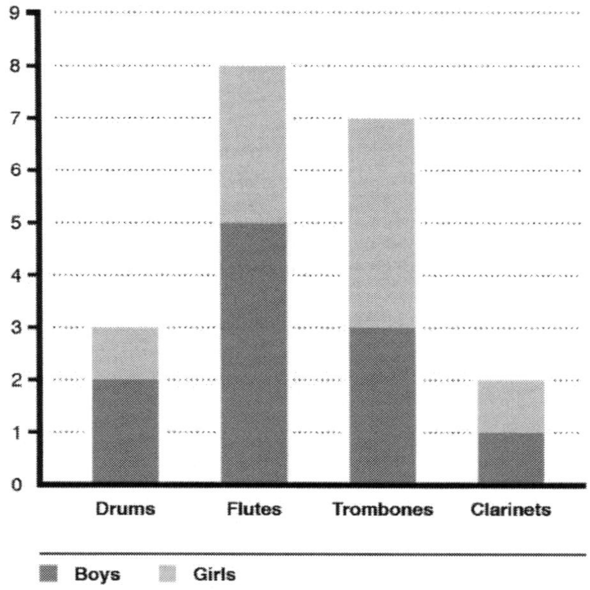

To determine how many boys play trombone, refer to the darker portion of the trombone bar, resulting in 3 students.

A **scatterplot** is another way to represent paired data. It uses Cartesian coordinates, like a line graph, meaning it has both a horizontal and vertical axis. Each data point is represented as a dot on the graph. The dots are never connected with a line. For example, the following is a scatterplot showing the connection between people's ages and heights.

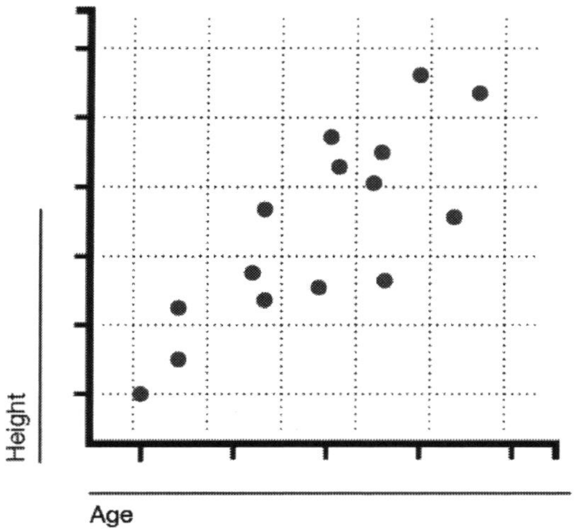

A scatterplot, also known as a **scattergram**, can be used to predict another value and to see if an association, known as a **correlation,** exists between a set of data. If the data resembles a straight line, the data is **associated.** The following is an example of a scatterplot in which the data does not seem to have an association:

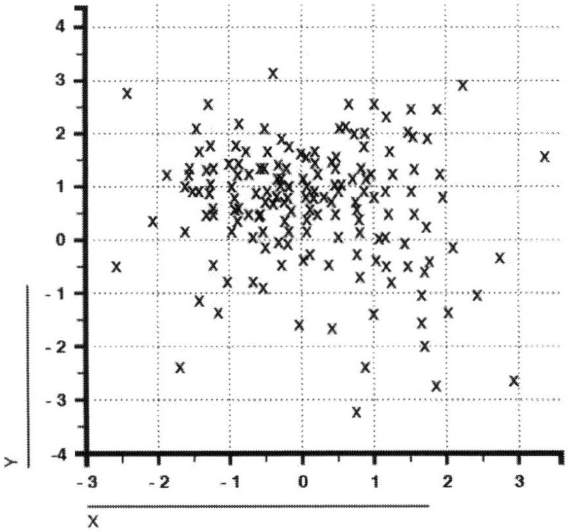

Sets of numbers and other similarly organized data can also be represented graphically. Venn diagrams are a common way to do so. A **Venn diagram** represents each set of data as a circle. The circles overlap, showing that each set of data is overlapping. A Venn diagram is also known as a **logic diagram** because it visualizes all possible logical combinations between two sets. Common elements of two sets are represented by the area of overlap. The following is an example of a Venn diagram of two sets A and B:

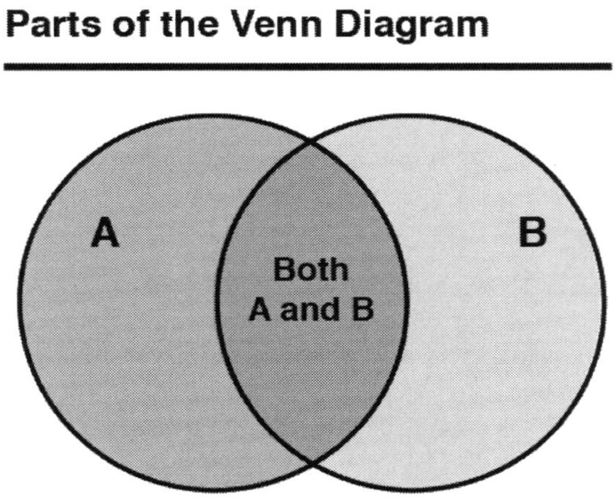

Another name for the area of overlap is the **intersection.** The intersection of A and B, $A \cap B$, contains all elements that are in both sets A and B. The **union** of A and B, $A \cup B$, contains all elements in both sets A and B. Finally, the **complement** of $A \cup B$ is equal to all elements that are not in either set A or set B. These elements are placed outside of the circles.

The following is an example of a Venn diagram in which 30 students were surveyed asking which type of siblings they had: brothers, sisters, or both. Ten students only had a brother, seven students only had a sister, and five had both a brother and a sister. Therefore, five is the intersection, represented by the section where the circles overlap. Two students did not have a brother or a sister. Two is therefore the complement and is placed outside of the circles.

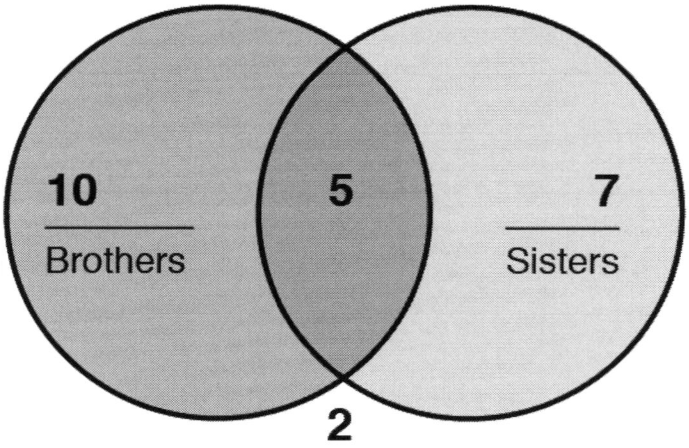

Venn diagrams can have more than two sets of data. The more circles, the more logical combinations are represented by the overlapping. The following is a Venn diagram that represents a different situation. Now, there were 30 students surveyed about the color of their socks. The innermost region represents those students that have green, pink, and blue socks on (perhaps a striped pattern). Therefore, two students had all three colors on their socks. In this example, all students had at least one of the three colors on their socks, so no one exists in the complement.

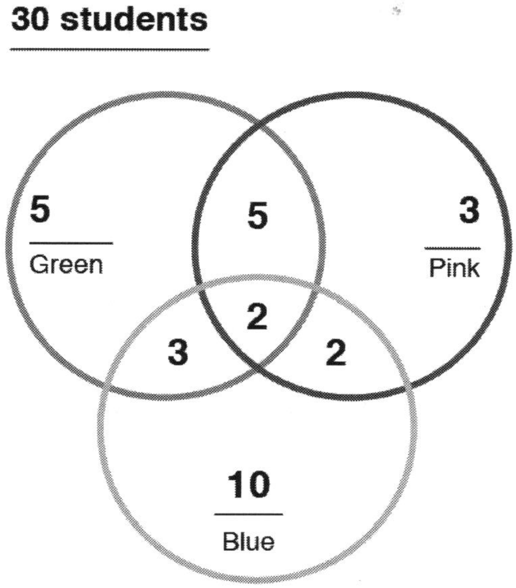

Venn diagrams are typically not drawn to scale; however, if they are, and if each circle's area is proportional to the amount of data it represents, then it is called an area-proportional Venn diagram.

Producing, Interpreting, Understanding, Evaluating, and Improving Models

Concrete models create a way of thinking about math that generates learning on a more permanent level. Memorizing abstract math formulas will not create a lasting effect on the brain. The following picture shows fractions represented by Lego blocks. By starting with the whole block of eight, it can be split into half, which is a four-block, and a fourth, which is a two-block. The one-eighth representation is a single block.

After splitting these up, addition and subtraction can be performed by adding or taking away parts of the blocks. Different combinations of fractions can be used to make a whole, or taken away to make various parts of a whole.

Using Colored Blocks to Model Functions

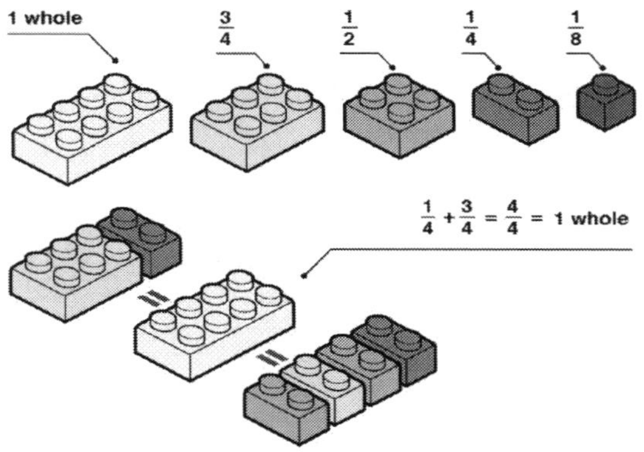

Multiplication can also be done using array models. The following picture shows a model of multiplying 3 times 4. **Arrays** are formed when the first factor is shown in a row. The second factor is shown in that number of columns. When the rectangle is formed, the blocks fill in to make a total, or the result of multiplication. The three rows and four columns show each factor and when the blocks are filled in, the total is twelve. Arrays are great ways to represent multiplication because they show each factor, and where the total comes from, with rows and columns.

Multiplication Array Model for 3 x 4

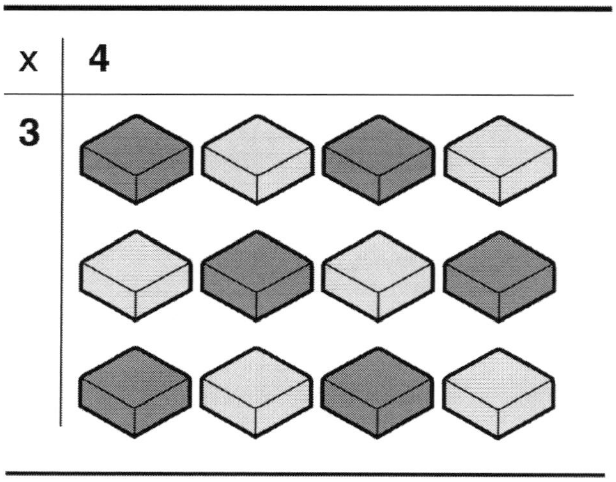

3 x 4 = 12

Another representation of fractions is shown below in the pie charts. Moving from whole numbers to part of numbers with fractions can be a concept that is difficult to grasp. Starting with a whole pie and splitting it into parts can be helpful with generating fractions. The first pie shows quarters or sections that are one-fourth because it is split into four parts. The second pie shows parts that equal one-fifth because it is split into five parts. Pies can also be used to add fractions. If $\frac{1}{5}$ and $\frac{1}{4}$ are being added, a common denominator must be found by splitting the pies into the same number of parts. The same number of parts can be found by determining the least common multiple. For 4 and 5, the least common multiple is 20. The pies can be split until there are 20 parts. The same portion of the pie can be shaded for each fraction and then added together to find the sum.

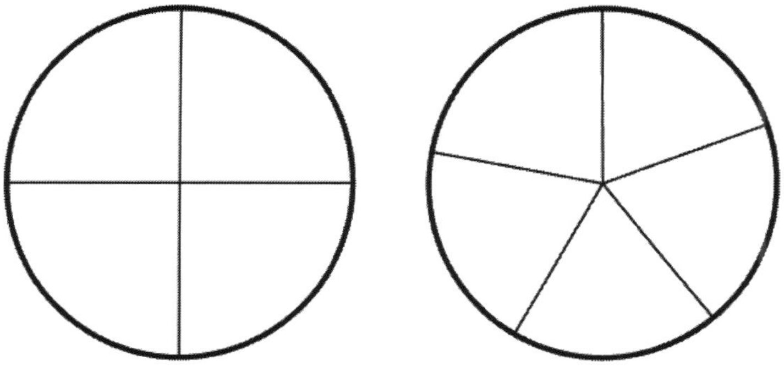

Illustrating and Explaining Multiplication and Division Problems Using Equations, Rectangular Arrays, and Area Models

Multiplication and division can be represented by equations. These equations show the numbers involved and the operation. For example, "eight multiplied by six equals forty-eight" is seen in the following equation:

$$8 \times 6 = 48$$

This operation can be modeled by rectangular arrays where one factor, 8, is the number of rows, and the other factor, 6, is the number of columns, as follows:

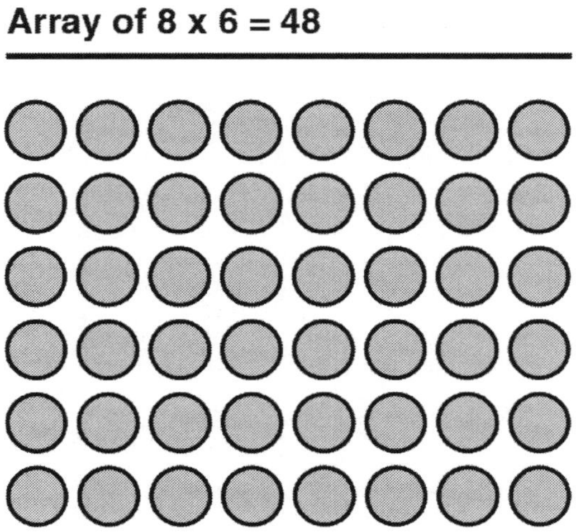

Rectangular arrays show what happens with the concept of multiplication. As one row of dots is drawn, that represents the first factor in the problem. Then the second factor is used to add the number of columns. The final model includes six rows of eight columns which results in forty-eight dots. These rectangular arrays show how multiplication of whole numbers will result in a number larger than the factors.

Division can also be represented by equations and area models. A division problem such as "twenty-four divided by three equals eight" can be written as the following equation:

$$24 \div 8 = 3$$

The object below shows an area model to represent the equation. As seen in the model, the whole box represents 24 and the 3 sections represent the division by 3. In more detail, there could be 24 dots written in the whole box and each box could have 8 dots in it.

Division shows how numbers can be divided into groups. For the example problem, it is asking how many numbers will be in each of the 3 groups that together make 24. The answer is 8 in each group.

Area Model of 24 ÷ 8 = 3

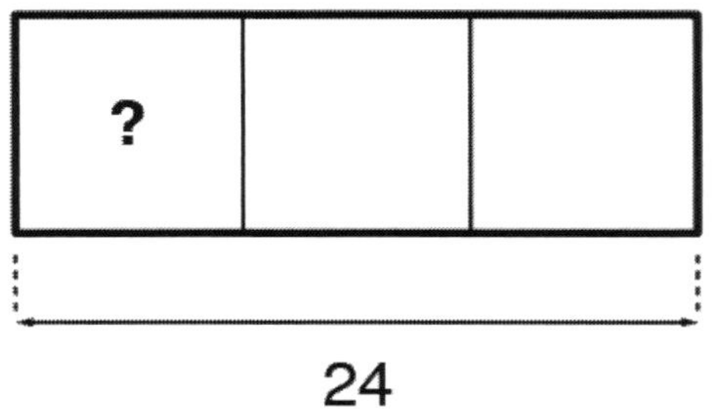

24 ÷ 3 = ?

Converting Within and Between Standard and Metric Systems

When working with dimensions, sometimes the given units don't match the formula, and conversions must be made. When performing operations with rational numbers, it might be helpful to round the numbers in the original problem to get a rough idea of what the answer should be. This system expands to three places above the base unit and three places below. These places correspond to prefixes that each signify a specific base of 10.

The following table shows the conversions:

kilo-	hecto-	deca-	base	deci-	centi-	milli-
1,000 times the base	100 times the base	10 times the base		1/10 times the base	1/100 times the base	1/1000 times the base

To convert between units within the metric system, values with a base ten can be multiplied. The decimal can also be moved in the direction of the new unit by the same number of zeros on the number. For example, 3 meters is equivalent to 0.003 kilometers. The decimal moved three places (the same number of zeros for kilo-) to the left (the same direction from base to kilo-). Three meters is also equivalent to 3,000 millimeters. The decimal is moved three places to the right because the prefix milli- is three places to the right of the base unit.

The English Standard system, which is used in the United States, uses the base units of foot for length, pound for weight, and gallon for liquid volume. Conversions within the English Standard system are not as easy as those within the metric system because the former does not use a base ten model. The following table shows the conversions within this system.

Length	Weight	Capacity
1 foot (ft) = 12 inches (in) 1 yard (yd) = 3 feet 1 mile (mi) = 5280 feet 1 mile = 1760 yards	1 pound (lb) = 16 ounces (oz) 1 ton = 2000 pounds	1 tablespoon (tbsp) = 3 teaspoons (tsp) 1 cup (c) = 16 tablespoons 1 cup = 8 fluid ounces (oz) 1 pint (pt) = 2 cups 1 quart (qt) = 2 pints 1 gallon (gal) = 4 quarts

When converting within the English Standard system, most calculations include a conversion to the base unit and then another to the desired unit. For example, take the following problem: $3 \text{ qt} = \_\_\_ \text{ c}$. There is no straight conversion from quarts to cups, so the first conversion is from quarts to pints. There are 2 pints in 1 quart, so there are 6 pints in 3 quarts. This conversion can be solved as a proportion:

$$\frac{3 \text{ qt}}{x} = \frac{1 \text{ qt}}{2 \text{ pt}}$$

It can also be observed as a ratio 2:1, expanded to 6:3. Then the 6 pints must be converted to cups. The ratio of pints to cups is 1:2, so the expanded ratio is 6:12. For 6 pints, the measurement is 12 cups. This problem can also be set up as one set of fractions to cancel out units. It begins with the given information and cancels out matching units on top and bottom to yield the answer. Consider the following expression:

$$\frac{3 \text{ qt}}{1} \times \frac{2 \text{ pt}}{1 \text{ qt}} \times \frac{2 \text{ c}}{1 \text{ pt}}$$

It's set up so that units on the top and bottom cancel each other out:

$$\frac{3 \not{\text{qt}}}{1} \times \frac{2 \not{\text{pt}}}{1 \not{\text{qt}}} \times \frac{2 \text{ c}}{1 \not{\text{pt}}}$$

The numbers can be calculated as 3 × 2 × 2 on the top and 1 on the bottom. It still yields an answer of 12 cups.

This process of setting up fractions and canceling out matching units can be used to convert between standard and metric systems. A few common equivalent conversions are $2.54 \text{ cm} = 1 \text{ in}$, $3.28 \text{ ft} = 1 \text{ m}$, and $2.205 \text{ lb} = 1 \text{ kg}$. Writing these as fractions allows them to be used in conversions. For the problem $5 \text{ meters} = \_\_\_$ ft, use the feet-to-meter conversion and start with the expression:

$$\frac{5 \text{ m}}{1} \times \frac{3.28 \text{ ft}}{1 \text{ m}}$$

The "meters" will cancel each other out, leaving "feet" as the final unit. Calculating the numbers yields 16.4 feet. This problem only required two fractions. Others may require longer expressions, but the underlying rule stays the same. When a unit in the numerator of a fraction matches a unit in the denominator, then they cancel each other out. Using this logic and the conversions given above, many units can be converted between and within the different systems.

The conversion between Fahrenheit and Celsius is found in a formula:

$$°\text{C} = (°\text{F} - 32) \times \frac{5}{9}$$

For example, to convert 78°F to Celsius, the given temperature would be entered into the formula:

$$°\text{C} = (78 - 32) \times \frac{5}{9}$$

Solving the equation, the temperature comes out to be 25.56°C. To convert in the other direction, the formula becomes:

$$°\text{F} = °\text{C} * \frac{9}{5} + 32$$

Remember the order of operations when calculating these conversions.

Solving Problems Involving Elapsed Time, Money, Length, Volume, and Mass

To solve problems, follow these steps: Identify the variables that are known, decide which equation should be used, substitute the numbers, and solve. To solve an equation for the amount of time that has elapsed since an event, use the equation T = L – E where T represents the elapsed time, L represents the later time, and E represents the earlier time. For example, the Minnesota Vikings have not appeared in the Super Bowl since 1976. If the year is now 2017, how long has it been since the Vikings were in the Super Bowl? The later time, L, is 2017, E = 1976 and the unknown is T.

Substituting these numbers, the equation is T = 2017 – 1976, and so T = 41. It has been 41 years since the Vikings have appeared in the Super Bowl. Questions involving total cost can be solved using the formula, C = I + T where C represents the total cost, I represents the cost of the item purchased, and T represents the tax amount. To find the length of a rectangle given the area = 32 square inches and width = 8 inches, the formula A = L × W can be used. Substitute 32 for A and substitute 8 for w, giving the equation 32 = L×8. This equation is solved by dividing both sides by 8 to find that the length of the rectangle is 4. The formula for volume of a rectangular prism is given by the equation:

$$V = L \times W \times H$$

If the length of a rectangular juice box is 4 centimeters, the width is 2 centimeters, and the height is 8 centimeters, what is the volume of this box? Substituting in the formula we find $V = 4 \times 2 \times 8$, so the volume is 64 cubic centimeters. In a similar fashion as those previously shown, the mass of an object can be calculated given the formula, $Mass = Density \times Volume$.

Measuring and Comparing Lengths of Objects Using Standard Tools

Lengths of objects can be measured using tools such as rulers, yard sticks, meter sticks, and tape measures. Typically, a ruler measures 12 inches, or one foot. For this reason, a ruler is normally used to measure lengths smaller than or just slightly more than 12 inches. Rulers may represent centimeters instead of inches. Some rulers have inches on one side and centimeters on the other. Be sure to recognize what units you are measuring in. The standard ruler measurements are divided into units of 1 inch and normally broken down to $\frac{1}{2}, \frac{1}{4}, \frac{1}{8}$, and even $\frac{1}{16}$ of an inch for more precise measurements. If measuring in centimeters, the centimeter is likely divided into tenths.

To measure the size of a picture, for purposes of buying a frame, a ruler is helpful. If the picture is very large, a yardstick, which measures 3 feet and normally is divided into feet and inches, might be useful. Using metric units, the meter stick measures 1 meter and is divided into 100 centimeters. To measure the size of a window in a home, either a yardstick or meter stick would work. To measure the size of a room, though, a tape measure would be the easiest tool to use. Tape measures can measure up to 10 feet, 25 feet, or more depending on the particular tape measure.

Comparing Relative Sizes of U.S. Customary Units and Metric Units

Measuring length in United States customary units is typically done using inches, feet, yards, and miles. When converting among these units, remember that 12 inches = 1 foot, 3 feet = 1 yard, and 5280 feet = 1 mile. Common customary units of weight are ounces and pounds. The conversion needed is 16 ounces = 1 pound. For customary units of volume ounces, cups, pints, quarts, and gallons are typically used. For conversions, use 8 ounces = 1 cup, 2 cups = 1 pint, 2 pints = 1 quart, and 4 quarts = 1 gallon. For measuring lengths in metric units, know that 100 centimeters = 1 meter, and 1000 meters = 1 kilometer. For metric units of measuring weights, grams and kilograms are often used. Know that 1000 grams = 1 kilogram when making conversions. For metric measures of volume, the most common units are milliliters and liters. Remember that 1000 milliliters = 1 liters.

ACT Math Practice Test #1

1. $3.4 + 2.35 + 4 =$
   a. 5.35
   b. 9.2
   c. 9.75
   d. 10.25
   e. 11.15

2. $5.88 \times 3.2 =$
   a. 18.816
   b. 16.44
   c. 20.352
   d. 17
   e. 19.25

3. $\frac{3}{25} =$
   a. 0.15
   b. 0.1
   c. 0.9
   d. 0.75
   e. 0.12

4. Which of the following is largest?
   a. 0.45
   b. 0.096
   c. 0.3
   d. 0.313
   e. 0.25

5. Which of the following is NOT a way to write 40 percent of *N*?
   a. $(0.4)N$
   b. $\frac{2}{5}N$
   c. $40N$
   d. $\frac{4N}{10}$
   e. $\frac{8N}{20}$

6. Which is closest to 17.8×9.9?
   a. 140
   b. 180
   c. 200
   d. 350
   e. 400

7. A student gets an 85% on a test with 20 questions. How many answers did the student solve correctly?
   a. 15
   b. 16
   c. 17
   d. 18
   e. 19

8. Four people split a bill. The first person pays for $\frac{1}{5}$, the second person pays for $\frac{1}{4}$, and the third person pays for $\frac{1}{3}$. What fraction of the bill does the fourth person pay?
   a. $\frac{13}{60}$
   b. $\frac{47}{60}$
   c. $\frac{1}{4}$
   d. $\frac{4}{15}$
   e. $\frac{1}{2}$

9. 6 is 30% of what number?
   a. 18
   b. 20
   c. 24
   d. 26
   e. 22

10. $3\frac{2}{3} - 1\frac{4}{5} =$
   a. $1\frac{13}{20}$
   b. $\frac{14}{15}$
   c. $2\frac{2}{3}$
   d. $\frac{4}{5}$
   e. $1\frac{13}{15}$

11. What is $\frac{420}{98}$ rounded to the nearest integer?
   a. 4
   b. 3
   c. 5
   d. 6
   e. 2

12. $4\frac{1}{3} + 3\frac{3}{4} =$

a. $6\frac{5}{12}$

b. $8\frac{1}{12}$

c. $8\frac{2}{3}$

d. $7\frac{7}{12}$

e. $9\frac{1}{12}$

13. Five of six numbers have a sum of 25. The average of all six numbers is 6. What is the sixth number?

a. 8
b. 10
c. 11
d. 12
e. 13

14. $52.3 \times 10^{-3} =$

a. 0.00523
b. 0.0523
c. 0.523
d. 523
e. 5,230

15. If $\frac{5}{2} \div \frac{1}{3} = n$, then n is between:

a. 5 and 7
b. 1 and 3
c. 9 and 11
d. 3 and 5
e. 7 and 9

16. A closet is filled with red, blue, and green shirts. If $\frac{1}{3}$ of the shirts are green and $\frac{2}{5}$ are red, what fraction of the shirts are blue?

a. $\frac{4}{15}$

b. $\frac{1}{5}$

c. $\frac{7}{15}$

d. $\frac{1}{2}$

e. $\frac{1}{3}$

17. Shawna buys $2\frac{1}{2}$ gallons of paint. If she uses $\frac{1}{3}$ of it on the first day, how much does she have left?

a. $1\frac{5}{6}$ gallons

b. $1\frac{1}{2}$ gallons

c. $1\frac{2}{3}$ gallons

d. 2 gallons

e. $2\frac{2}{3}$ gallons

18. If $6t + 4 = 16$, what is t?

a. 1
b. 2
c. 3
d. 4
e. 5

19. There are $4x + 1$ treats in each party favor bag. If a total of $60x + 15$ treats is distributed, how many bags are given out?

a. 15
b. 16
c. 20
d. 22
e. 24

20. In an office, there are 50 workers. A total of 60% of the workers are women, and the chances of a woman wearing a skirt is 50%. If no men wear skirts, how many workers are wearing skirts?

a. 12
b. 15
c. 16
d. 20
e. 21

21. If $2x + 6 = 20$, what is x?

a. 3
b. 4
c. 7
d. 9
e. 11

22. The variable y is directly proportional to x. If $y = 3$ when $x = 5$, then what is y when $x = 20$?

a. 10
b. 12
c. 14
d. 16
e. 18

23. Apples cost $2 each, while oranges cost $3 each. Maria purchased 10 fruits in total and spent $22. How many apples did she buy?

a. 5
b. 6
c. 7
d. 4
e. 8

24. What are the polynomial roots of $x^2 + x - 2$?

a. 1 and -2
b. -1 and 2
c. 2 and -2
d. 9 and 13
e. 3 and 9

25. What is the *y*-intercept of $y = x^{5/3} + (x - 3)(x + 1)$?

a. 3.5
b. 7.6
c. -3
d. -15.1
e. 3

26. Suppose $\frac{x+2}{x} = 2$. What is *x*?

a. -1
b. 0
c. 2
d. 4
e. -2

27. A rectangle has a length that is 5 feet longer than three times its width. If the perimeter is 90 feet, what is the length in feet?

a. 10
b. 20
c. 25
d. 35
e. 40

28. Five students take a test. The scores of the first four students are 80, 85, 75, and 60. If the median score is 80, which of the following could NOT be the score of the fifth student?

a. 60
b. 80
c. 85
d. 100
e. 90

29. Change $3\frac{3}{5}$ to a decimal.

a. 3.6
b. 4.67
c. 5.3
d. 0.28
e. 6.3

30. If a car can travel 300 miles in 4 hours, how far can it go in an hour and a half?

a. 100 miles
b. 112.5 miles
c. 135.5 miles
d. 150 miles
e. 155 miles

31. Which measure for the center of a small sample set is most affected by outliers?

a. Mean
b. Median
c. Mode
d. Range
e. None of these

32. Given the value of a given stock at monthly intervals, which graph should be used to best represent the trend of the stock?

a. Box plot
b. Line plot
c. Scatter plot
d. Circle graph
e. Line graph

33. What is the probability of randomly picking the winner and runner-up from a race of 4 horses and distinguishing which is the winner?

a. $\frac{1}{4}$
b. $\frac{1}{2}$
c. $\frac{1}{16}$
d. $\frac{1}{12}$
e. $\frac{1}{3}$

34. Which of the following could be used in the classroom to show $\frac{3}{7} < \frac{5}{6}$ is a true statement?

a. A bar graph
b. A number line
c. An area model
d. Base 10 blocks
e. A line graph

35. Add $103{,}678 + 487$
   a. 103,191
   b. 103,550
   c. 104,265
   d. 104,165
   e. 105,143

36. Add $1.001 + 5.629$
   a. 4.472
   b. 4.628
   c. 5.630
   d. 6.245
   e. 6.630

37. Add $143.77 + 5.2$
   a. 138.57
   b. 148.97
   c. 138.97
   d. 148.57
   e. 149.087

38. What is the next number in the following series: $1, 3, 6, 10, 15, 21, \dots$?
   a. 26
   b. 27
   c. 28
   d. 29
   e. 30

39. Add and express in reduced form $\frac{14}{33} + \frac{10}{11}$.
   a. $\frac{2}{11}$
   b. $\frac{6}{11}$
   c. $\frac{4}{3}$
   d. $\frac{44}{33}$
   e. $\frac{3}{4}$

40. 32 is 25% of what number?
   a. 64
   b. 128
   c. 12.65
   d. 8
   e. 155

41. Subtract $112{,}076 - 1{,}243$.
   a. 109,398
   b. 113,319
   c. 113,833
   d. 110,833
   e. 111,485

42. Carey bought 184 pounds of fertilizer to use on her lawn. Each segment of her lawn required $11\frac{1}{2}$ pounds of fertilizer to do a sufficient job. If a student were asked to determine how many segments could be fertilized with the amount purchased, what operation would be necessary to solve this problem?
   a. Multiplication
   b. Division
   c. Addition
   d. Subtraction
   e. Exponents

43. Subtract $701.1 - 52.33$.
   a. 753.43
   b. 648.77
   c. 652.77
   d. 638.43
   e. 786.34

44. Which of the following expressions best exemplifies the additive and subtractive identity?
   a. $5 + 2 - 0 = 5 + 2 + 0$
   b. $6 + x = 6 - 6$
   c. $9 - 9 = 0$
   d. $8 + 2 = 10$
   e. $7 + 8 = 15$

45. Which of the following is an equivalent measurement for 1.3 cm?
   a. 0.13 m
   b. 13 m
   c. 0.13 mm
   d. 0.013 mm
   e. 0.013 m

46. Divide $1{,}015 \div 1.4$.
   a. 7,250
   b. 0.725
   c. 7.25
   d. 725
   e. 575

47. At the store, Jan spends $90 on apples and oranges. Apples cost $1 each and oranges cost $2 each. If Jan buys the same number of apples as oranges, how many oranges did she buy?
  a. 20
  b. 25
  c. 30
  d. 35
  e. 40

48. Multiply 12.4×0.2.
  a. 12.6
  b. 2.48
  c. 12.48
  d. 2.6
  e. 4.28

49. Multiply $1{,}987 \times 0.05$.
  a. 9.935
  b. 99.35
  c. 993.5
  d. 999.35
  e. 93.95

50. A clock reads 5:00 am. What is the measure of the angle formed by the two hands of that clock?
  a. 300 degrees
  b. 150 degrees
  c. 75 degrees
  d. 210 degrees
  e. 285 degrees

51. Divide, express with a remainder $1{,}202 \div 44$.
  a. $27\frac{2}{7}$
  b. $2\frac{7}{22}$
  c. $7\frac{2}{7}$
  d. $27\frac{7}{22}$
  e. $22\frac{7}{22}$

52. What is the volume of a box with rectangular sides 5 feet long, 6 feet wide, and 3 feet high?
  a. 60 cubic feet
  b. 75 cubic feet
  c. 80 cubic feet
  d. 14 cubic feet
  e. 90 cubic feet

53. Divide $702 \div 2.6$.
    a. 27
    b. 207
    c. 2.7
    d. 270
    e. 2700

54. A train traveling 50 miles per hour takes a trip lasting 3 hours. If a map has a scale of 1 inch per 10 miles, how many inches apart are the train's starting point and ending point on the map?
    a. 14
    b. 12
    c. 13
    d. 15
    e. 16

55. A traveler takes an hour to drive to a museum, spends 3 hours and 30 minutes there, and takes half an hour to drive home. What percentage of his or her time was spent driving?
    a. 15%
    b. 30%
    c. 40%
    d. 60%
    e. 70%

56. A truck is carrying three cylindrical barrels. Their bases have a diameter of 2 feet, and they have a height of 3 feet. What is the total volume of the three barrels in cubic feet?
    a. 3π
    b. 9π
    c. 12π
    d. 15π
    e. 18π

57. Greg buys a $10 lunch with 5% sales tax. He leaves a $2 tip after his bill. How much money does he spend?
    a. $12.50
    b. $12
    c. $13
    d. $13.25
    e. $13.50

58. Add $5{,}089 + 10{,}323$
    a. 15,402
    b. 15,412
    c. 5,234
    d. 15,234
    e. 16,001

59. A teacher is showing students how to evaluate $5 \times 6 + 4 \div 2 - 1$. Which operation should be completed first?
   a. Multiplication
   b. Addition
   c. Division
   d. Subtraction
   e. Exponent

60. What is the definition of a factor of the number 36?
   a. A number that can be divided by 36 and have no remainder
   b. A number that can be added to 36 with no remainder
   c. A prime number that is multiplied times 36
   d. A number that 36 can be divided by and have no remainder
   e. A number that 36 can be multiplied by and have no remainder

Answer Explanations for Practice Test #1

1. C: The decimal points are lined up, with zeroes put in as needed. Then, the numbers are added just like integers:

$$\begin{array}{r} 3.40 \\ 2.35 \\ +\underline{4.00} \\ 9.75 \end{array}$$

2. A: This problem can be multiplied as 588×32, except at the end, the decimal point needs to be moved three places to the left. Performing the multiplication will give 18,816, and moving the decimal place over three places results in 18.816.

3. E: The fraction is converted so that the denominator is 100 by multiplying the numerator and denominator by 4, to get:

$$\frac{3}{25} = \frac{12}{100}$$

Dividing a number by 100 just moves the decimal point two places to the left, with a result of 0.12.

4. A: Figure out which is largest by looking at the first non-zero digits. Choice *B*'s first non-zero digit is in the hundredths place. The other four all have non-zero digits in the tenths place, so it must be *A*, *C*, *D*, or *E*. Of these, *A* has the largest first non-zero digit.

5. C: $40N$ would be 4000% of N. It's possible to check that each of the others is actually 40% of N.

6. B: Instead of multiplying these out, the product can be estimated by using $18 \times 10 = 180$. The error here should be lower than 15, since it is rounded to the nearest integer, and the numbers add to something less than 30.

7. C: 85% of a number means multiplying that number by 0.85. So:

$$0.85 \times 20 = \frac{85}{100} \times \frac{20}{1}$$

which can be simplified to:

$$\frac{17}{20} \times \frac{20}{1} = 17$$

8. A: To find the fraction of the bill that the first three people pay, the fractions need to be added, which means finding the common denominator. The common denominator will be 60.

$$\frac{1}{5} + \frac{1}{4} + \frac{1}{3} = \frac{12}{60} + \frac{15}{60} + \frac{20}{60} = \frac{47}{60}$$

The remainder of the bill is:

$$1 - \frac{47}{60} = \frac{60}{60} - \frac{47}{60} = \frac{13}{60}$$

9. B: 30% is 3/10. The number itself must be 10/3 of 6, or:

$$\frac{10}{3} \times 6 = 10 \times 2 = 20$$

10. E: These numbers to improper fractions: $\frac{11}{3} - \frac{9}{5}$. Take 15 as a common denominator:

$$\frac{11}{3} - \frac{9}{5} =: \frac{55}{15} - \frac{27}{15} = \frac{28}{15} = 1\frac{13}{15}$$

11. A: Dividing by 98 can be approximated by dividing by 100, which would mean shifting the decimal point of the numerator to the left by 2. The result is 4.2 and rounds to 4.

12. B: $4\frac{1}{3} + 3\frac{3}{4} = 4 + 3 + \frac{1}{3} + \frac{3}{4} = 7 + \frac{1}{3} + \frac{3}{4}$

Adding the fractions gives:

$$\frac{1}{3} + \frac{3}{4} = \frac{4}{12} + \frac{9}{12} = \frac{13}{12} = 1 + \frac{1}{12}$$

Thus:

$$7 + \frac{1}{3} + \frac{3}{4} = 7 + 1 + \frac{1}{12} = 8\frac{1}{12}$$

13. C: The average is calculated by adding all six numbers, then dividing by 6. The first five numbers have a sum of 25. If the total divided by 6 is equal to 6, then the total itself must be 36. The sixth number must be 36 – 25 = 11.

14. B: Multiplying by 10^{-3} means moving the decimal point three places to the left, putting in zeroes as necessary.

15. E:

$$\frac{5}{2} \div \frac{1}{3} = \frac{5}{2} \times \frac{3}{1} = \frac{15}{2} = 7.5$$

16. A: The total fraction taken up by green and red shirts will be:

$$\frac{1}{3} + \frac{2}{5} = \frac{5}{15} + \frac{6}{15} = \frac{11}{15}$$

The remaining fraction is:

$$1 - \frac{11}{15} = \frac{15}{15} - \frac{11}{15} = \frac{4}{15}$$

17. C: If she has used 1/3 of the paint, she has 2/3 remaining. $2\frac{1}{2}$ gallons are the same as $\frac{5}{2}$ gallons.

The calculation is:

$$\frac{2}{3} \times \frac{5}{2} = \frac{5}{3} = 1\frac{2}{3} \text{ gallons}$$

18. B: First, subtract 4 from each side. This yields $6t = 12$. Now, divide both sides by 6 to obtain $t = 2$.

19. A: Each bag contributes $4x + 1$ treats. The total treats will be in the form $4nx + n$ where n is the total number of bags. The total is in the form $60x + 15$, from which it is known $n = 15$.

20. B: If 60% of 50 workers are women, then there are 30 women working in the office. If half of them are wearing skirts, then that means 15 women wear skirts. Since none of the men wear skirts, this means there are 15 people wearing skirts.

21. C: Begin by subtracting 6 from both sides to get $2x = 14$. Dividing both sides by 2 results in $x = 7$.

22. B: To be directly proportional means that $y = mx$. If x is changed from 5 to 20, the value of x is multiplied by 4. Applying the same rule to the y-value, also multiply the value of y by 4. Therefore, $y = 12$.

23. E: Let a be the number of apples and o the number of oranges. Then, the total cost is:

$$2a + 3o = 22$$

while it also known that $a + o = 10$. Using the knowledge of systems of equations, cancel the o variables by multiplying the second equation by -3.

This makes the equation:

$$-3a - 3o = -30$$

Adding this to the first equation, the b values cancel to get $-a = -8$, which simplifies to $a = 8$.

24. A: Finding the roots means finding the values of x when y is zero. The quadratic formula could be used, but in this case it is possible to factor by hand, since the numbers -1 and 2 add to 1 and multiply to -2. So, factor:

$$x^2 + x - 2 = (x - 1)(x + 2) = 0$$

then set each factor equal to zero. Solving for each value gives the values $x = 1$ and $x = -2$.

25. C: To find the y-intercept, substitute zero for x, which gives us:

$$y = 0^{5/3} + (0 - 3)(0 + 1) = 0 + (-3)(1) = -3$$

26. C: Multiply both sides by x to get $x + 2 = 2x$, which simplifies to $-x = -2$, or $x = 2$.

27. D: Denote the width as w and the length as l. Then, $l = 3w + 5$. The perimeter is $2w + 2l = 90$. Substituting the first expression for l into the second equation yields:

$$2(3w + 5) + 2w = 90$$

$$6w + 10 + 2w = 90$$

$$8w = 80$$

$$w = 10$$

Putting this into the first equation, it yields:

$$l = 3(10) + 5 = 35$$

28. A: Lining up the given scores provides the following list: 60, 75, 80, 85, and one unknown. Because the median needs to be 80, it means 80 must be the middle data point out of these five. Therefore, the unknown data point must be the fourth or fifth data point, meaning it must be greater than or equal to 80. The only answer that fails to meet this condition is 60.

29. A: 3.6

Divide 3 by 5 to get 0.6 and add that to the whole number 3, to get 3.6. An alternative is to incorporate the whole number 3 earlier on by creating an improper fraction: $\frac{18}{5}$. Then dividing 18 by 5 to get 3.6.

30. B: 300 miles in 4 hours is $\frac{300}{4}$ = 75 miles per hour. In 1.5 hours, the car will go 1.5×75 miles, or 112.5 miles.

31. A: Mean. An outlier is a data value that is either far above or far below the majority of values in a sample set. The mean is the average of all the values in the set. In a small sample set, a very high or very low number could drastically change the average of the data points. Outliers will have no more of an effect on the median (the middle value when arranged from lowest to highest) than any other value above or below the median. If the same outlier does not repeat, outliers will have no effect on the mode (value that repeats most often).

32. E: Line graph. The scenario involves data consisting of two variables, month and stock value. Box plots display data consisting of values for one variable. Therefore, a box plot is not an appropriate choice. Both line plots and circle graphs are used to display frequencies within categorical data. Neither can be used for the given scenario. Scatter plots compare the values of two variables to see if there are any patterns present. Line graphs display two numerical variables on a coordinate grid and show trends among the variables.

33. D: $\frac{1}{12}$. The probability of picking the winner of the race is $\frac{1}{4}$, or:

$$\left(\frac{number\ of\ favorable\ outcomes}{number\ of\ total\ outcomes}\right)$$

Assuming the winner was picked on the first selection, three horses remain from which to choose the runner-up (these are dependent events). Therefore, the probability of picking the runner-up is $\frac{1}{3}$. To determine the probability of multiple events, the probability of each event is multiplied:

$$\frac{1}{4} \times \frac{1}{3} = \frac{1}{12}$$

34. B: This inequality can be seen with the use of a number line. $\frac{3}{7}$ is close to $\frac{1}{2}$.

$\frac{5}{6}$ is close to 1, but less than 1, and $\frac{8}{7}$ is greater than 1. Therefore, $\frac{3}{7}$ is less than $\frac{5}{6}$.

35. D: 104,165

Set up the problem and add each column, starting on the far right (ones). Add, carrying anything over 9 into the next column to the left. Solve from right to left.

36. E: 6.630

Set up the problem, with the larger number on top and numbers lined up at the decimal. Add, carrying anything over 9 into the next column to the left. Solve from right to left.

37. B: 148.97

Set up the problem, with the larger number on top and numbers lined up at the decimal. Insert 0 in any blank spots to the right of the decimal as placeholders. Add, carrying anything over 9 into the next column to the left.

38. C: Each number in the sequence is adding one more than the difference between the previous two.

For example, $10 - 6 = 4, 4 + 1 = 5$.

Therefore, the next number after 10 is $10 + 5 = 15$.

Going forward, $21 - 15 = 6, 6 + 1 = 7$. The next number is $21 + 7 = 28$.

Therefore, the difference between numbers is the set of whole numbers starting at 2: 2, 3, 4, 5, 6, 7....

39. C: $\frac{4}{3}$

Set up the problem and find a common denominator for both fractions:

$$\frac{14}{33} + \frac{10}{11}$$

Multiply each fraction across by 1 to convert to a common denominator:

$$\frac{14}{33} \times \frac{1}{1} + \frac{10}{11} \times \frac{3}{3}$$

Once over the same denominator, add across the top. The total is over the common denominator:

$$\frac{14 + 30}{33} = \frac{44}{33}$$

Reduce by dividing both numerator and denominator by 11:

$$\frac{44 \div 11}{33 \div 11} = \frac{4}{3}$$

40. B: 128

This question involves the percent formula:

$$\frac{32}{x} = \frac{25}{100}$$

We multiply the diagonal numbers, 32 and 100, to get 3,200. Dividing by the remaining number, 25, gives us 128.

The percent formula does not have to be used for a question like this. Since 25% is ¼ of 100, you know that 32 needs to be multiplied by 4, which yields 128.

41. D: 110,833

Set up the problem, with the larger number on top. Begin subtracting with the far right column (ones). Borrow 10 from the column to the left, when necessary.

42. B: This is a division problem because the original amount needs to be split up into equal amounts. The mixed number $11\frac{1}{2}$ should be converted to an improper fraction first:

$$11\frac{1}{2} = \frac{(11 \times 2) + 1}{2} = \frac{23}{2}$$

Carey needs to determine how many times $\frac{23}{2}$ goes into 184. This is a division problem:

$$184 \div \frac{23}{2} = ?$$

The fraction can be flipped, and the problem turns into the multiplication:

$$184 \times \frac{2}{23} = \frac{368}{23}$$

This improper fraction can be simplified into 16 because $368 \div 23 = 16$. The answer is 16 lawn segments.

43. B: 648.77

Set up the problem, with the larger number on top and numbers lined up at the decimal. Insert 0 in any blank spots to the right of the decimal as placeholders. Begin subtracting with the far right column. Borrow 10 from the column to the left, when necessary.

44. A: The additive and subtractive identity is 0. When added or subtracted to any number, 0 does not change the original number.

45. E: 100 cm is equal to 1 m. 1.3 divided by 100 is 0.013. Therefore, 1.3 cm is equal to 0.013 m. Because 1 cm is equal to 10 mm, 1.3 cm is equal to 13 mm.

46. D: 725

Set up the division problem.

1.	4	1	0	1	5

Move the decimal over one place to the right in both numbers.

1	4	1	0	1	5	0

14 does not go into 1 or 10 but does go into 101 so start there.

				7	2	5
1	4	1	0	1	5	0
		-	9	8		
				3	5	
			-	2	8	
					7	0
				-	7	0
						0

The result is 725.

47. C: One apple/orange pair costs \$3 total. Therefore, Jan bought $\frac{90}{3} = 30$ total pairs, and hence, she bought 30 oranges.

48. B: 2.48

Set up the problem, with the larger number on top. Multiply as if there are no decimal places. Add the answer rows together. Count the number of decimal places that were in the original numbers ($1 + 1 = 2$).

Place the decimal 2 places to the right for the final solution.

49. B: 99.35

Set up the problem, with the larger number on top. Multiply as if there are no decimal places. Add the answer rows together. Count the number of decimal places that were in the original numbers (2).

Place the decimal in that many spots from the right for the final solution.

50. B: Each hour on the clock represents 30 degrees. For example, 3:00 represents a right angle. Therefore, 5:00 represents 150 degrees.

51. D: $27\frac{7}{22}$

Set up the division problem.

$$44\overline{)\,1\ 2\ 0\ 2}$$

44 does not go into 1 or 12 but will go into 120 so start there.

				2	7
4	4	1	2	0	2
		-	8	8	
			3	2	2
		-	3	0	8
				1	4

The answer is $27\frac{14}{44}$.

Reduce the fraction for the final answer.

$$27\frac{7}{22}$$

52. E: The formula for the volume of a box with rectangular sides is the length times width times height, so $5 \times 6 \times 3 = 90$ cubic feet.

53. D: 270

Set up the division problem.

$$2.6\overline{)\,7\ 0\ 2}$$

Move the decimal over one place to the right in both numbers.

$$26\overline{)\,7\ 0\ 2\ 0}$$

26 does not go into 7 but does go into 70 so start there.

				2	7	0
2	6		7	0	2	0
		-	5	2		
			1	8	2	
		-	1	8	2	
					0	

The result is 270.

54. D: First, the train's journey in the real world is $3 \times 50 = 150$ miles. On the map, 1 inch corresponds to 10 miles, so there is $\frac{150}{10} = 15$ inches on the map.

55. B: The total trip time is 1 + 3.5 + 0.5 = 5 hours. The total time driving is 1 + 0.5 = 1.5 hours. So, the fraction of time spent driving is 1.5/5 or 3/10. To get the percentage, convert this to a fraction out of 100.

The numerator and denominator are multiplied by 10, with a result of 30/100. The percentage is the numerator in a fraction out of 100, so 30%.

56. B: The formula for the volume of a cylinder is $\pi r^2 h$, where r is the radius and h is the height. The diameter is twice the radius, so these barrels have a radius of 1 foot. That means each barrel has a volume of $\pi \times 1^2 \times 3 = 3$ cubic feet. Since there are three of them, the total is $3 \times 3\pi = 9\pi$ cubic feet.

57. A: The tip is not taxed, so he pays 5% tax only on the $10. 5% of $10 is $0.05 \times 10 = \$0.50$. Add up $\$10 + \$2 + \$0.50$ to get $12.50.

58. B: 15,412

Set up the problem and add each column, starting on the far right (ones). Add, carrying anything over 9 into the next column to the left. Solve from right to left.

59. A: Using the order of operations, multiplication and division are computed first from left to right. Multiplication is on the left; therefore, multiplication should be performed first.

60. D: A factor of 36 is any number that can be divided into 36 and have no remainder. $36 = 36 \times 1, 18 \times 2, 9 \times 4, \text{and } 6 \times 6$. Therefore, it has 7 unique factors: 36, 18, 9, 6, 4, 2, and 1.

ACT Math Practice Test #2

1. Which of the following numbers has the greatest value?
   a. 1.4378
   b. 1.07548
   c. 1.43592
   d. 0.89409
   e. 0.94739

2. The value of 6×12 is the same as:
   a. $2 \times 4 \times 4 \times 2$
   b. $7 \times 4 \times 3$
   c. $6 \times 6 \times 3$
   d. $3 \times 3 \times 4 \times 2$
   e. $5 \times 9 \times 8$

3. This chart indicates how many sales of CDs, vinyl records, and MP3 downloads occurred over the last year. Approximately what percentage of the total sales was from CDs?

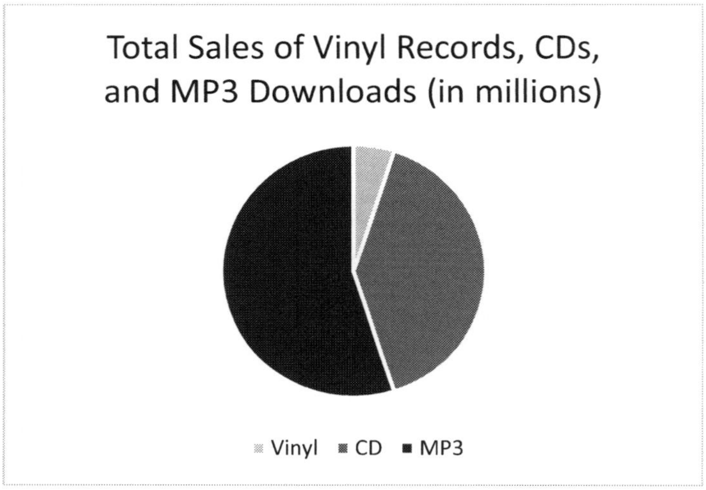

   a. 55%
   b. 25%
   c. 40%
   d. 5%
   e. 15%

4. Alan currently weighs 200 pounds, but he wants to lose weight to get down to 175 pounds. What is this difference in kilograms? (1 pound is approximately equal to 0.45 kilograms.)
   a. 9 kg
   b. 18.55 kg
   c. 78.75 kg
   d. 90 kg
   e. 11.25 kg

5. Johnny earns $2334.50 from his job each month. He pays $1437 for monthly expenses. Johnny is planning a vacation in 3 months that he estimates will cost $1750 total. How much will Johnny have left over from three months of saving once he pays for his vacation?

   a. $948.50
   b. $584.50
   c. $852.50
   d. $942.50
   e. $742.50

6. Solve the following:

$$(\sqrt{36} \times \sqrt{16}) - 3^2$$

   a. 30
   b. 21
   c. 15
   d. 13
   e. 25

7. In Jim's school, there are 3 girls for every 2 boys. There are 650 students in total. Using this information, how many students are girls?

   a. 260
   b. 130
   c. 65
   d. 390
   e. 410

8. Kimberley earns $10 an hour babysitting, and after 10 p.m., she earns $12 an hour, with the amount paid being rounded to the nearest hour accordingly. On her last job, she worked from 5:30 p.m. to 11 p.m. In total, how much did Kimberley earn on her last job?

   a. $45
   b. $57
   c. $62
   d. $42
   e. $53

9. Arrange the following numbers from least to greatest value:

   $0.85, \frac{4}{5}, \frac{2}{3}, \frac{91}{100}$

   a. $0.85, \frac{4}{5}, \frac{2}{3}, \frac{91}{100}$

   b. $\frac{4}{5}, 0.85, \frac{91}{100}, \frac{2}{3}$

   c. $\frac{2}{3}, \frac{4}{5}, 0.85, \frac{91}{100}$

   d. $0.85, \frac{91}{100}, \frac{4}{5}, \frac{2}{3}$

e. $\frac{4}{5}, \frac{91}{100}, \frac{2}{3}, 0.85$

10. Keith's bakery had 252 customers go through its doors last week. This week, that number increased to 378. Express this increase as a percentage.
a. 26%
b. 50%
c. 35%
d. 12%
e. 18%

11. Simplify the following expression:

$$4\frac{2}{3} - 3\frac{4}{9}$$

a. $1\frac{1}{3}$

b. $1\frac{3}{8}$

c. 1

d. $1\frac{2}{3}$

e. $1\frac{2}{9}$

12. Jessica buys 10 cans of paint. Red paint costs $1 per can and blue paint costs $2 per can. In total, she spends $16. How many red cans did she buy?
a. 2
b. 3
c. 4
d. 5
e. 6

13. Six people apply to work for Janice's company, but she only needs four workers. How many different groups of four employees can Janice choose?
a. 6
b. 10
c. 15
d. 36
e. 42

14. Which of the following is equivalent to the value of the digit 3 in the number 792.134?
a. 3×10

b. 3×100

c. $\frac{3}{10}$

d. $\frac{3}{100}$

e. $100 \div 3$

15. In the following expression, which operation should be completed first? $5 \times 6 + (5 + 4) \div 2 - 1$.
    a. Multiplication
    b. Addition
    c. Division
    d. Parentheses
    e. Exponents

16. How will the number 847.89632 be written if rounded to the nearest hundredth?
    a. 847.90
    b. 900
    c. 847.89
    d. 847.896
    e. 850

17. The perimeter of a 6-sided polygon is 56 cm. The length of three of the sides are 9 cm each. The length of two other sides are 8 cm each. What is the length of the missing side?
    a. 11 cm
    b. 12 cm
    c. 13 cm
    d. 10 cm
    e. 9 cm

18. Which of the following is a mixed number?
    a. $16\frac{1}{2}$
    b. 16
    c. $\frac{16}{3}$
    d. $\frac{1}{4}$
    e. $\frac{90}{80}$

19. Change 9.3 to a fraction.
    a. $9\frac{3}{7}$
    b. $\frac{903}{1000}$
    c. $\frac{9.03}{100}$
    d. $3\frac{9}{10}$
    e. $9\frac{3}{10}$

20. What is the value of b in this equation?

$$5b - 4 = 2b + 17$$

a. 13
b. 24
c. 7
d. 21
e. 15

21. What is 39% of 164?
a. 63.96
b. 23.78
c. 6,396
d. 2.38
e. .0987

22. Katie works at a clothing company and sold 192 shirts over the weekend. One third of the shirts that were sold were patterned, and the rest were solid. Which mathematical expression would calculate the number of solid shirts Katie sold over the weekend?

a. $192 \times \frac{1}{3}$

b. $192 \div \frac{1}{3}$

c. $192 \times (1 - \frac{1}{3})$

d. $192 \div 3$

e. $192 \times (1 + \frac{1}{3})$

23. Which four-sided shape is always a rectangle?
a. Rhombus
b. Square
c. Parallelogram
d. Quadrilateral
e. Trapezoid

24. A rectangle was formed out of pipe cleaner. Its length was $\frac{1}{2}$ ft, and its width was $\frac{11}{2}$ inches. What is its area in square inches?

a. $\frac{11}{4}$ inch^2

b. $\frac{11}{2}$ inch^2

c. 22 inches^2

d. $27\frac{1}{3}$ inches^2

e. 33 inches^2

25. How will $\frac{4}{5}$ be written as a percent?

a. 40 percent
b. 125 percent
c. 90 percent
d. 80 percent
e. 25 percent

26. If Danny takes 48 minutes to walk 3 miles, how long should it take him to walk 5 miles maintaining the same speed?

a. 32 min
b. 64 min
c. 80 min
d. 96 min
e. 105 min

27. A solution needs 5 ml of saline for every 8 ml of medicine given. How much saline is needed for 45 ml of medicine?

a. $\frac{225}{8}$ ml

b. 72 ml

c. 28 ml

d. $\frac{45}{8}$ ml

e. 84 ml

28. What unit of volume is used to describe the following 3-dimensional shape?

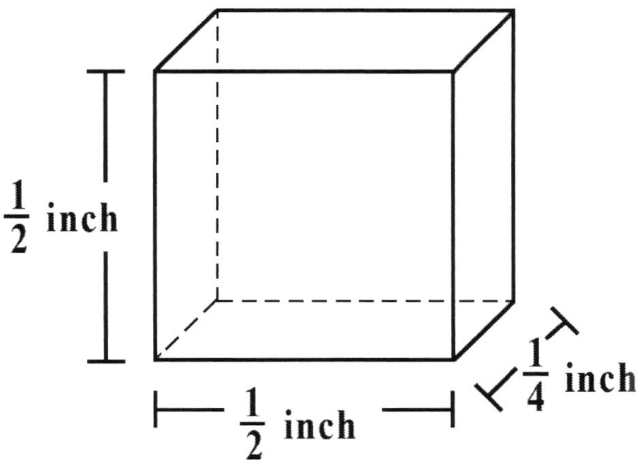

a. Square inches
b. Inches
c. Cubic inches
d. Squares
e. Cubic feet

29. Which common denominator would be used in order to evaluate $\frac{2}{3} + \frac{4}{5}$?
    a. 15
    b. 3
    c. 5
    d. 10
    e. 12

30. The diameter of a circle measures 5.75 centimeters. What tool could be used to draw such a circle?
    a. Ruler
    b. Meter stick
    c. Compass
    d. Yard stick
    e. Protractor

31. A piggy bank contains 12 dollars' worth of nickels. A nickel weighs 5 grams, and the empty piggy bank weighs 1050 grams. What is the total weight of the full piggy bank?
    a. 1,110 grams
    b. 1,200 grams
    c. 2,150 grams
    d. 2,250 grams
    e. 2,500 grams

32. Last year, the New York City area received approximately $27\frac{3}{4}$ inches of snow. The Denver area received approximately 3 times as much snow as New York City. How much snow fell in Denver?
    a. $71\frac{3}{4}$ inches
    b. $27\frac{1}{4}$ inches
    c. $89\frac{1}{4}$ inches
    d. $83\frac{1}{4}$ inches
    e. $64\frac{3}{4}$ inches

33. Which of the following would be an instance in which ordinal numbers are used?
    a. Katie scored a 9 out of 10 on her quiz.
    b. Matthew finished second in the spelling bee.
    c. Jacob missed one day of school last month.
    d. Kim was 5 minutes late to school this morning.
    e. Katrina lost 10 lbs. over the summer.

34. How will the following number be written in standard form: $(1 \times 10^4) + (3 \times 10^3) + (7 \times 10^1) + (8 \times 10^0)$
    a. 137
    b. 13,780
    c. 1,378
    d. 13,078

e. 3,780

35. What is the area of the regular hexagon shown below?

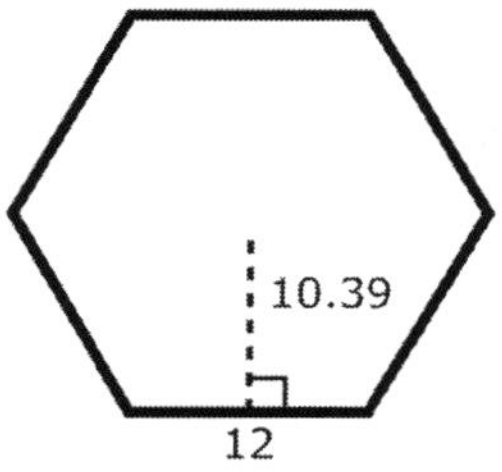

a. 72
b. 124.68
c. 374.04
d. 748.08
e. 676.79

36. The area of a given rectangle is 24 square centimeters. If the measure of each side is multiplied by 3, what is the area of the new figure?
a. 48 cm^2
b. 72 cm^2
c. 111 cm^2
d. 13,824 cm^2
e. 216 cm^2

37. Which of the following is the definition of a prime number?
a. A number that factors only into itself and 1
b. A number greater than one that factors only into itself and 1
c. A number less than 10
d. A number divisible by 10
e. Any number that's greater than 1

38. Add and express in reduced form $\frac{5}{12} + \frac{4}{9}$
a. $\frac{9}{17}$
b. $\frac{1}{3}$
c. $\frac{31}{36}$
d. $\frac{3}{5}$
e. $\frac{15}{29}$

39. Which of the following is the correct order of operations that could be used on a difficult math problem that contained grouping symbols?

a. Parentheses, Exponents, Multiplication, Division, Addition, Subtraction
b. Exponents, Parentheses, Multiplication, Division, Addition, Subtraction
c. Parentheses, Exponents, Addition, Multiplication, Division, Subtraction
d. Parentheses, Exponents, Division, Addition, Subtraction, Multiplication
e. Multiplication, Parentheses, Exponents, Division, Subtraction, Addition

40. Convert $\frac{5}{8}$ to a decimal.

a. 0.62
b. 1.05
c. 0.63
d. 1.60
e. 1.25

41. Subtract $9{,}576 - 891$.

a. 10,467
b. 9,685
c. 8,325
d. 8,685
e. 7,564

42. If a teacher was showing a class how to round 245.2678 to the nearest thousandth, which place value would be used to decide whether to round up or round down?

a. Ten-thousandth
b. Thousandth
c. Hundredth
d. Thousand
e. Hundred

43. Subtract $50.888 - 13.091$.

a. 63.799
b. 63.979
c. 37.979
d. 37.797
e. 36.697

44. Students should line up decimal places within the given numbers before performing which of the following?

a. Multiplication
b. Division
c. Subtraction
d. Exponents
e. Parentheses

45. Subtract and express in reduced form $\frac{23}{24} - \frac{1}{6}$.
   a. $\frac{22}{18}$
   b. $\frac{11}{9}$
   c. $\frac{19}{24}$
   d. $\frac{4}{5}$
   e. $\frac{1}{2}$

46. Subtract and express in reduced form $\frac{43}{45} - \frac{11}{15}$.
   a. $\frac{10}{45}$
   b. $\frac{16}{15}$
   c. $\frac{32}{30}$
   d. $\frac{5}{9}$
   e. $\frac{2}{9}$

47. Change 0.56 to a fraction.
   a. $\frac{5.6}{100}$
   b. $\frac{14}{25}$
   c. $\frac{56}{1000}$
   d. $\frac{56}{10}$
   e. $\frac{100}{56}$

48. Multiply $13,114 \times 191$.
   a. 2,504,774
   b. 250,477
   c. 150,474
   d. 2,514,774
   e. 1,759,492

49. Marty wishes to save $150 over a 4-day period. How much must Marty save each day on average?
   a. $37.50
   b. $35
   c. $45.50
   d. $41
   e. $43.50

50. A teacher cuts a pie into 6 equal pieces and takes one away. What topic would she be introducing to the class by using such a visual?
   a. Decimals
   b. Addition
   c. Subtraction
   d. Fractions
   e. Exponents

51. Multiply and reduce $\frac{15}{23} \times \frac{54}{127}$.
   a. $\frac{810}{2,921}$
   b. $\frac{81}{292}$
   c. $\frac{69}{150}$
   d. $\frac{810}{2929}$
   e. $\frac{2921}{810}$

52. Which of the following represent one hundred eighty-two billion, thirty-six thousand, four hundred twenty-one and three hundred fifty-six thousandths?
   a. 182,036,421.356
   b. 182,036,421.0356
   c. 182,000,036,421.0356
   d. 182,000,036,421.356
   e. 182,360,000,421.356

53. Divide, express with a remainder $188 \div 16$.
   a. $1\frac{3}{4}$
   b. $111\frac{3}{4}$
   c. $10\frac{3}{4}$
   d. $11\frac{3}{4}$
   e. $3\frac{4}{11}$

54. What other operation could be utilized to teach the process of dividing 9453 by 24 besides division?
   a. Multiplication
   b. Addition
   c. Exponents
   d. Subtraction
   e. Parentheses

55. Bernard can make \$80 per day. If he needs to make \$300 and only works full days, how many days will this take?
   a. 6
   b. 3
   c. 5
   d. 2
   e. 4

56. A couple buys a house for \$150,000. They sell it for \$165,000. By what percentage did the house's value increase?
   a. 18%
   b. 13%
   c. 15%
   d. 12%
   e. 10%

57. What operation is repeated to evaluate an expression involving an exponent?
   a. Addition
   b. Multiplication
   c. Division
   d. Subtraction
   e. Parentheses

58. Which of the following formulas would correctly calculate the perimeter of a legal-sized piece of paper that is 14 inches long and $8\frac{1}{2}$ inches wide?
   a. $P = 14 + 8\frac{1}{2}$
   b. $P = 14 + 8\frac{1}{2} + 14 + 8\frac{1}{2}$
   c. $P = 14 \times 8\frac{1}{2}$
   d. $P = 14 \times \frac{17}{2}$
   e. $P = 14 - \frac{17}{2}$

59. Which of the following are units that would be taught in a lecture covering the metric system?
   a. Inches, feet, miles, pounds
   b. Millimeters, centimeters, meters, pounds
   c. Kilograms, grams, kilometers, meters
   d. Teaspoons, tablespoons, ounces
   e. Stone, quart, gallon, foot, yard

60. Which important mathematical property is shown in the expression: $(7 \times 3) \times 2 = 7 \times (3 \times 2)$?
   a. Distributive property
   b. Commutative property
   c. Additive inverse
   d. Associative property
   e. Multiplicative Identity property

Answer Explanations for Practice Test #2

1. A: Compare each numeral after the decimal point to figure out which overall number is greatest. In answers *A* (1.43785) and *C* (1.43592), both have the same tenths (4) and hundredths (3). However, the thousandths is greater in answer *A* (7), so *A* has the greatest value overall.

2. D: By grouping the four numbers in the answer into factors of the two numbers of the question (6 and 12), it can be determined that:

$$(3 \times 2) \times (4 \times 3) = 6 \times 12$$

Alternatively, each of the answer choices could be prime factored or multiplied out and compared to the original value.

6×12 has a value of 72 and a prime factorization of $2^3 \times 3^2$.

The answer choices respectively have values of 64, 84, 108, 72, and 360, so Choice *D* is correct.

3. C: The sum total percentage of a pie chart must equal 100%. Since the CD sales take up less than half of the chart and more than a quarter (25%), it can be determined to be 40% overall. This can also be measured with a protractor. The angle of a circle is 360°. Since 25% of 360° would be 90° and 50% would be 180°, the angle percentage of CD sales falls in between; therefore, it would be Choice *C*.

4. B: Using the conversion rate, multiply the projected weight loss of 25 lb. by 0.45 $\frac{kg}{lb}$ to get the amount in kilograms (11.25 kg).

5. D: First, subtract $1437 from $2334.50 to find Johnny's monthly savings; this equals $897.50. Then, multiply this amount by 3 to find out how much he will have in three months before he pays for his vacation; this equals $2692.50. Finally, subtract the cost of the vacation ($1750) from this amount to find how much Johnny will have left: $942.50.

6. C: Follow the *order of operations* in order to solve this problem. Solve the parentheses first, and then follow the remainder as usual:

$$(6 \times 4) - 9$$

This equals 24 – 9 or 15, answer *C*.

7. D: Three girls for every two boys can be expressed as a ratio: 3:2. This can be visualized as splitting the school into 5 groups: 3 girl groups and 2 boy groups. The number of students which are in each group can be found by dividing the total number of students by 5:

$$\frac{650 \text{ students}}{5 \text{ groups}} = \frac{130 \text{ students}}{\text{group}}$$

To find the total number of girls, multiply the number of students per group (130) by the number of girl groups in the school (3). This equals 390, Choice *D*.

8. C: Kimberley worked 4.5 hours at the rate of $10/h and 1 hour at the rate of $12/h. The problem states that her pay is rounded to the nearest hour, so the 4.5 hours would round up to 5 hours at the rate of $10/h:

$$(5h)(\$10/h) + (1h)(\$12/h) = \$50 + \$12 = \$62$$

9. C: The first step is to depict each number using decimals. $\frac{91}{100} = 0.91$

Dividing the numerator by denominator of $\frac{4}{5}$ to convert it to a decimal yields 0.80, while $\frac{2}{3}$ becomes 0.66 recurring. Rearrange each expression in ascending order, as found in Choice *C*.

10. B: First, calculate the difference between the larger value and the smaller value:

$$378 - 252 = 126$$

To calculate this difference as a percentage of the original value, and thus calculate the percentage *increase*, divide 126 by 252, then multiply by 100 to reach the percentage 50%, Choice *B*.

11. E: Simplify each mixed number of the problem into a fraction by multiplying the denominator by the whole number and adding the numerator:

$$\frac{14}{3} - \frac{31}{9}$$

Since the first denominator is a multiple of the second, simplify it further by multiplying both the numerator and denominator of the first expression by 3 so that the denominators of the fractions are equal:

$$\frac{42}{9} - \frac{31}{9} = \frac{11}{9}$$

Simplifying this further, divide the numerator 11 by the denominator 9; this leaves 1 with a remainder of 2. To write this as a mixed number, place the remainder over the denominator, resulting in $1\frac{2}{9}$.

12. C: We are trying to find x, the number of red cans. The equation can be set up like this:

$$x + 2(10 - x) = 16$$

The left x is actually multiplied by $1, the price per red can. Since we know Jessica bought 10 total cans, $10 - x$ is the number blue cans that she bought. We multiply the number of blue cans by $2, the price per blue can.

That should all equal $16, the total amount of money that Jessica spent. Working that out gives us:

$$x + 20 - 2x = 16$$

$$20 - x = 16$$

$$x = 4$$

13. C: Janice will be choosing 4 employees out of a set of 6 applicants, so this will be given by the choice function. The following equation shows the choice function worked out:

$$\binom{6}{4} = \frac{6!}{4!\,(6-4)!} = \frac{6!}{4!\,(2)!}$$

$$\frac{6 \times 5 \times 4 \times 3 \times 2 \times 1}{4 \times 3 \times 2 \times 1 \times 2 \times 1} = \frac{6 \times 5}{2} = 15$$

14. D: $\frac{3}{100}$. Each digit to the left of the decimal point represents a higher multiple of 10 and each digit to the right of the decimal point represents a quotient of a higher multiple of 10 for the divisor.

The first digit to the right of the decimal point is equal to the value ÷ 10. The second digit to the right of the decimal point is equal to the value ÷ (10×10), or the value ÷ 100.

15. D: Using the order of operations, items inside of parentheses are sorted out first. Therefore, the teacher should resolve the parentheses first. In this expression, multiplication and division are computed next from left to right, followed by addition and subtraction.

16. A: 847.90.

The hundredths place value is located two digits to the right of the decimal point (the digit 9 in the original number). The digit to the right of the place value is examined to decide whether to round up or keep the digit. In this case, the digit 6 is 5 or greater so the hundredth place is rounded up. When rounding up, if the digit to be increased is a 9, the digit to its left is increased by one and the digit in the desired place value is made a zero. Therefore, the number is rounded to 847.90.

17. C: Perimeter is found by calculating the sum of all sides of the polygon.

$9 + 9 + 9 + 8 + 8 + s = 56$, where s is the missing side length. Therefore, 43 plus the missing side length is equal to 56. The missing side length is 13 cm.

18. A: $16\frac{1}{2}$. A mixed number contains both a whole number and either a fraction or a decimal. Therefore, the mixed number is $16\frac{1}{2}$.

19. E: $9\frac{3}{10}$. To convert a decimal to a fraction, remember that any number to the left of the decimal point will be a whole number. Then, since 0.3 goes to the tenths place, it can be placed over 10.

20. C: To solve for the value of *b*, both sides of the equation need to be equalized.

Start by cancelling out the lower value of -4 by adding 4 to both sides:

$$5b - 4 = 2b + 17$$
$$5b - 4 + 4 = 2b + 17 + 4$$
$$5b = 2b + 21$$

The variable *b* is the same on each side, so subtract the lower 2b from each side:

$$5b = 2b + 21$$
$$5b - 2b = 2b + 21 - 2b$$
$$3b = 21$$

Then divide both sides by 3 to get the value of *b*:

$$3b = 21$$
$$\frac{3b}{3} = \frac{21}{3}$$
$$b = 7$$

21. A: 63.96

This question involves the percent formula. Since we're beginning with a percent, also known as a number over 100, we'll put 39 on the right side of the equation:

$$\frac{x}{164} = \frac{39}{100}$$

Now, multiply 164 and 39 to get 6,396, which then needs to be divided by 100.

$$6{,}396 \div 100 = 63.96$$

22. C: $\frac{1}{3}$ of the shirts sold were patterned. Therefore, $1 - \frac{1}{3} = \frac{2}{3}$ of the shirts sold were solid. Anytime "of" a quantity appears in a word problem, multiplication needs to be used. Therefore, $192 \times \frac{2}{3} = 192 \times \frac{2}{3} = \frac{384}{3} = 128$ solid shirts were sold. The entire expression is $192 \times \left(1 - \frac{1}{3}\right)$.

23. B: A rectangle is a specific type of parallelogram. It has 4 right angles. A square is a rhombus that has 4 right angles. Therefore, a square is always a rectangle because it has two sets of parallel lines and 4 right angles.

24. E: Recall the formula for area, area $= \text{length} \times \text{width}$. The answer must be in square inches, so all values must be converted to inches. Half of a foot is equal to 6 inches. Therefore, the area of the rectangle is equal to:

$$6 \text{ in} \times \frac{11}{2} \text{in} = \frac{66}{2} \text{in}^2 = 33 \text{ in}^2$$

25. D: 80 percent. To convert a fraction to a percent, the fraction is first converted to a decimal. To do so, the numerator is divided by the denominator: $4 \div 5 = 0.8$. To convert a decimal to a percent, the number is multiplied by 100: $0.8 \times 10 = 80\%$.

26. C: 80 min. To solve the problem, a proportion is written consisting of ratios comparing distance and time. One way to set up the proportion is:

$$\frac{3}{48} = \frac{5}{x} \left(\frac{distance}{time} = \frac{distance}{time}\right)$$

where x represents the unknown value of time. To solve a proportion, the ratios are cross-multiplied:

$$(3)(x) = (5)(48) \rightarrow 3x = 240$$

The equation is solved by isolating the variable, or dividing by 3 on both sides, to produce $x = 80$.

27. A: Every 8 ml of medicine requires 5 ml. The 45 ml first needs to be split into portions of 8 ml. This results in $\frac{45}{8}$ portions. Each portion requires 5 ml. Therefore:

$$\frac{45}{8} \times 5 = 45 \times \frac{5}{8} = \frac{225}{8}$$

ml is necessary.

28. C: Volume of this 3-dimensional figure is calculated using length x width x height. Each measure of length is in inches. Therefore, the answer would be labeled in cubic inches.

29. A: A common denominator must be found. The least common denominator is 15 because it has both 5 and 3 as factors. The fractions must be rewritten using 15 as the denominator.

30. C: A compass is a tool that can be used to draw a circle. The compass would be drawn by using the length of the radius, which is half of the diameter.

31. D: A dollar contains 20 nickels. Therefore, if there are 12 dollars' worth of nickels, there are $12 \times 20 = 240$ nickels.

Each nickel weighs 5 grams. Therefore, the weight of the nickels is $240 \times 5 = 1{,}200$ grams.

Adding in the weight of the empty piggy bank, the filled bank weighs 2,250 grams.

32. D: To find Denver's total snowfall, 3 must be multiplied times $27\frac{3}{4}$. In order to easily do this, the mixed number should be converted into an improper fraction.

$$27\frac{3}{4} = \frac{27 \times 4 + 3}{4} = \frac{111}{4}$$

Therefore, Denver had approximately $\frac{3 \times 111}{4} = \frac{333}{4}$ inches of snow. The improper fraction can be converted back into a mixed number through division.

$$\frac{333}{4} = 83\frac{1}{4} \text{inches}$$

33. B: Ordinal numbers represent a ranking. Placing second in a competition is a ranking among the other participants of the spelling bee.

34. D: 13,078. The power of 10 by which a digit is multiplied corresponds with the number of zeros following the digit when expressing its value in standard form. Therefore:

$$(1 \times 10^4) + (3 \times 10^3) + (7 \times 10^1) + (8 \times 10^0)$$

$$10{,}000 + 3{,}000 + 70 + 8 = 13{,}078$$

35. C: 374.04. The formula for finding the area of a regular polygon is $A = \frac{1}{2} \times a \times P$ where *a* is the length of the apothem (from the center to any side at a right angle), and P is the perimeter of the figure. The apothem *a* is given as 10.39, and the perimeter can be found by multiplying the length of one side by the number of sides (since the polygon is regular): $P = 12 \times 6 \rightarrow P = 72$. To find the area, substitute the values for *a* and *P* into the formula

$$A = \frac{1}{2} \times a \times P$$

$$A = \frac{1}{2} \times (10.39) \times (72) \rightarrow A = 374.04$$

36. E: 216 cm^2. Because area is a two-dimensional measurement, the dimensions are multiplied by a scale that is squared to determine the scale of the corresponding areas. The dimensions of the rectangle are multiplied by a scale of 3. Therefore, the area is multiplied by a scale of 3^2 (which is equal to 9):

$$24\ cm^2 \times 9 = 216\ cm^2$$

37. B: A number is prime because its only factors are itself and 1. Positive numbers (greater than one) can be prime numbers.

38. C: $\frac{31}{36}$

Set up the problem and find a common denominator for both fractions.

$$\frac{5}{12} + \frac{4}{9}$$

Multiply each fraction across by 1 to convert to a common denominator.

$$\frac{5}{12} \times \frac{3}{3} + \frac{4}{9} \times \frac{4}{4}$$

Once over the same denominator, add across the top. The total is over the common denominator.

$$\frac{15 + 16}{36} = \frac{31}{36}$$

39. A: Order of operations follows PEMDAS—Parentheses, Exponents, Multiplication and Division from left to right, and Addition and Subtraction from left to right.

40. C: 0.63

Divide 5 by 8, which results in 0.63.

41. D: 8,685

Set up the problem, with the larger number on top. Begin subtracting with the far-right column (ones). Borrow 10 from the column to the left, when necessary.

42. A: The place value to the right of the thousandth place, which would be the ten-thousandth place, is what gets used. The value in the thousandth place is 7. The number in the place value to its right is greater than 4, so the 7 gets bumped up to 8. Everything to its right turns to a zero, to get 245.2680. The zero is dropped because it is part of the decimal.

43. D: 37.797

Set up the problem, larger number on top and numbers lined up at the decimal. Begin subtracting with the far-right column. Borrow 10 from the column to the left, when necessary.

44. C: Numbers should be lined up by decimal places before subtraction is performed. This is because subtraction is performed within each place value. The other operations, such as multiplication, division, and exponents (which is a form of multiplication), involve ignoring the decimal places at first and then including them at the end.

45. C: $\frac{19}{24}$

Set up the problem and find a common denominator for both fractions.

$$\frac{23}{24} - \frac{1}{6}$$

Multiply each fraction across by 1 to convert to a common denominator.

$$\frac{23}{24} \times \frac{1}{1} - \frac{1}{6} \times \frac{4}{4}$$

Once over the same denominator, subtract across the top.

$$\frac{23 - 4}{24} = \frac{19}{24}$$

46. E: $\frac{2}{9}$

Set up the problem and find a common denominator for both fractions.

$$\frac{43}{45} - \frac{11}{15}$$

Multiply each fraction across by 1 to convert to a common denominator.

$$\frac{43}{45} \times \frac{1}{1} - \frac{11}{15} \times \frac{3}{3}$$

Once over the same denominator, subtract across the top.

$$\frac{43 - 33}{45} = \frac{10}{45}$$

Reduce.

$$\frac{10 \div 5}{45 \div 5} = \frac{2}{9}$$

47. B: $\frac{14}{25}$

Since 0.56 goes to the hundredths place, it can be placed over 100:

$$\frac{56}{100}$$

Essentially, the way we got there is by multiplying the numerator and denominator by 100:

$$\frac{0.56}{1} \times \frac{100}{100} = \frac{56}{100}$$

Then, the fraction can be simplified down to $\frac{14}{25}$:

$$\frac{56}{100} \div \frac{4}{4} = \frac{14}{25}$$

48. A: 2,504,774

Line up the numbers (the number with the most digits on top) to multiply. Begin with the right column on top and the right column on bottom.

Move one column left on top and multiply by the far-right column on the bottom. Remember to add the carry over after you multiply. Continue that pattern for each of the numbers on the top row.

Starting on the far-right column on top repeat this pattern for the next number left on the bottom. Write the answers below the first line of answers; remember to begin with a zero placeholder. Continue for each number in the top row.

Starting on the far-right column on top, repeat this pattern for the next number left on the bottom. Write the answers below the first line of answers. Remember to begin with zero placeholders.

Once completed, ensure the answer rows are lined up correctly, then add.

49. A: The first step is to divide up $150 into four equal parts. $\frac{150}{4}$ is 37.5, so she needs to save an average of $37.50 per day.

50. D: The teacher would be introducing fractions. If a pie was cut into 6 pieces, each piece would represent $\frac{1}{6}$ of the pie. If one piece was taken away, $\frac{5}{6}$ of the pie would be left over.

51. A: $\frac{810}{2921}$

Line up the fractions.

$$\frac{15}{23} \times \frac{54}{127}$$

Multiply across the top and across the bottom:

$$\frac{15 \times 54}{23 \times 127} = \frac{810}{2921}$$

52. D: There are no millions, so the millions period consists of all zeros. 182 is in the billions period, 36 is in the thousands period, 421 is in the hundreds period, and 356 is the decimal.

53. D: $11\frac{3}{4}$

Set up the division problem.

$$16 \overline{)188}$$

16 does not go into 1 but does go into 18 so start there.

$$\begin{array}{r} 11 \\ 16 \overline{)188} \\ -\underline{16} \\ 28 \\ -\underline{16} \\ 12 \end{array}$$

The result is $11\frac{12}{16}$

Reduce the fraction for the final answer.

$$11\frac{3}{4}$$

54. D: Division can be computed as a repetition of subtraction problems by subtracting multiples of 24.

55. E: The number of days can be found by taking the total amount Bernard needs to make and dividing it by the amount he earns per day:

$$\frac{300}{80} = \frac{30}{8} = \frac{15}{4} = 3.75$$

But Bernard is only working full days, so he will need to work 4 days, since 3 days is not a sufficient amount of time.

56. E: The value went up by:

$$\$165{,}000 - \$150{,}000 = \$15{,}000$$

Out of $150,000, this is $\frac{15{,}000}{150{,}000} = \frac{1}{10}$. Convert this to having a denominator of 100, the result is $\frac{10}{100}$ or 10%.

57. B: A number raised to an exponent is a compressed form of multiplication.

For example, $10^3 = 10 \times 10 \times 10$.

58. B: The perimeter of a rectangle is the sum of all four sides.

Therefore, the answer is:

$$P = 14 + 8\frac{1}{2} + 14 + 8\frac{1}{2}$$

$$14 + 14 + 8 + \frac{1}{2} + 8 + \frac{1}{2} = 45 \text{ square inches}$$

59. C: Kilograms, grams, kilometers, and meters. Inches, pounds, and baking measurements, such as tablespoons, are not part of the metric system. Kilograms, grams, kilometers, and meters are part of the metric system.

60. D: It shows the associative property of multiplication. The order of multiplication does not matter, and the grouping symbols do not change the final result once the expression is evaluated.

ACT Math Practice Test #3

1. At the beginning of the day, Xavier has 20 apples. At lunch, he meets his sister Emma and gives her half of his apples. After lunch, he stops by his neighbor Jim's house and gives him 6 of his apples. He then uses $\frac{3}{4}$ of his remaining apples to make an apple pie for dessert at dinner. At the end of the day, how many apples does Xavier have left?

   a. 4
   b. 6
   c. 2
   d. 1
   e. 3

2. What is the product of two irrational numbers?

   a. Irrational
   b. Rational
   c. Irrational or rational
   d. Complex and imaginary
   e. Imaginary

3. The graph shows the position of a car over a 10-second time interval. Which of the following is the correct interpretation of the graph for the interval 1 to 3 seconds?

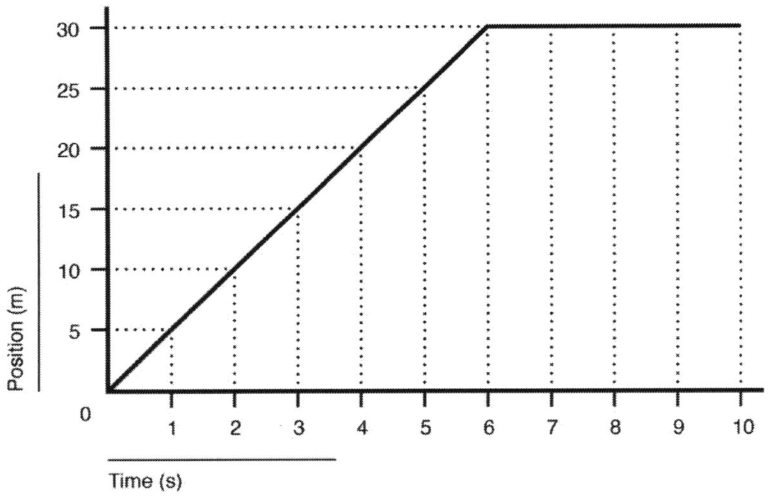

   a. The car remains in the same position.
   b. The car is traveling at a speed of 5 m/s.
   c. The car is traveling up a hill.
   d. The car is traveling at 5 mph.
   e. The car accelerates at a rate of 5 m/s.

4. Being as specific as possible, how is the number -4 classified?
   a. Real, rational, integer, whole, natural
   b. Real, rational, integer, natural
   c. Real, rational, integer
   d. Real, irrational, complex
   e. Real, irrational, whole

5. After a 20% discount, Frank purchased a new refrigerator for $850. How much did he save from the original price?
   a. $170
   b. $212.50
   c. $105.75
   d. $200
   e. $187.50

6. Store brand coffee beans cost $1.23 per pound. A local coffee bean roaster charges $1.98 per 1 ½ pounds. How much more would 5 pounds from the local roaster cost than 5 pounds of the store brand?
   a. $0.55
   b. $1.55
   c. $1.45
   d. $0.45
   e. $0.35

7. What is the solution to the following problem in decimal form?

$$\frac{3}{5} \times \frac{7}{10} \div \frac{1}{2}$$

   a. 0.042
   b. 84%
   c. 0.84
   d. 0.42
   e. 42%

8. Dwayne has received the following scores on his math tests: 78, 92, 83, and 97. What score must Dwayne get on his next math test to have an overall average of 90?
   a. 89
   b. 98
   c. 95
   d. 94
   e. 100

9. What are all the factors of 12?
   a. 12, 24, 36
   b. 1, 2, 4, 6, 12
   c. 12, 24, 36, 48
   d. 1, 2, 3, 4, 6, 12
   e. 0, 1, 12

10. Which of the following augmented matrices represents the system of equations below?

$$2x - 3y + z = -5$$
$$4x - y - 2z = -7$$
$$-x + 2z = -1$$

a. $\begin{bmatrix} 2 & -3 & 1 & 5 \\ 4 & -1 & 0 & -7 \\ -1 & 0 & 2 & 1 \end{bmatrix}$

b. $\begin{bmatrix} 2 & 4 & -1 \\ -3 & -1 & 0 \\ 1 & -2 & 2 \\ -5 & -7 & -1 \end{bmatrix}$

c. $\begin{bmatrix} 2 & 4 & -1 & -5 \\ -3 & -1 & 0 & -7 \\ 2 & -2 & 2 & -1 \end{bmatrix}$

d. $\begin{bmatrix} 2 & -3 & 1 \\ 4 & -1 & -2 \\ -1 & 0 & 2 \end{bmatrix}$

e. $\begin{bmatrix} 2 & -3 & 1 & -5 \\ 4 & -1 & -2 & -7 \\ -1 & 0 & 2 & -1 \end{bmatrix}$

11. What are the zeros of the function: $f(x) = x^3 + 4x^2 + 4x$?

a. -2
b. 0, -2
c. 2
d. 0, 2
e. 0

12. If $g(x) = x^3 - 3x^2 - 2x + 6$ and $f(x) = 2$, then what is $g(f(x))$?

a. -26
b. 6
c. $2x^3 - 6x^2 - 4x + 12$
d. -2
e. $2x^2 - 6$

13. What is the solution to the following system of equations?

$$x^2 - 2x + y = 8$$
$$x - y = -2$$

a. $(-2, 3)$
b. There is no solution.
c. $(-2, 0)\ (1, 3)$
d. $(-2, 0)\ (3, 5)$
e. $(2, 0)\ (-1, 3)$

14. Which of the following is the result after simplifying the expression: $(7n + 3n^3 + 3) + (8n + 5n^3 + 2n^4)$?
 a. $9n^4 + 15n - 2$
 b. $2n^4 + 5n^3 + 15n - 2$
 c. $9n^4 + 8n^3 + 15n$
 d. $2n^4 + 8n^3 + 15n + 3$
 e. $2n^4 + 5n^3 + 15n - 3$

15. What is the product of the following expression?

$$(4x - 8)(5x^2 + x + 6)$$

 a. $20x^3 - 36x^2 + 16x - 48$
 b. $6x^3 - 41x^2 + 12x + 15$
 c. $20x^4 + 11x^2 - 37x - 12$
 d. $2x^3 - 11x^2 - 32x + 20$
 e. $20x^3 - 40x^2 + 24x - 48$

16. How could the following equation be factored to find the zeros?

$$y = x^3 - 3x^2 - 4x$$

 a. $0 = x^2(x - 4), x = 0, 4$
 b. $0 = 3x(x + 1)(x + 4), x = 0, -1, -4$
 c. $0 = x(x + 1)(x + 6), x = 0, -1, -6$
 d. $0 = 3x(x + 1)(x - 4), x = 0, 1, -4$
 e. $0 = x(x + 1)(x - 4), x = 0, -1, 4$

17. The hospital has a nurse to patient ratio of 1:25. If there is a maximum of 325 patients admitted at a time, how many nurses are there?
 a. 13 nurses
 b. 25 nurses
 c. 325 nurses
 d. 12 nurses
 e. 5 nurses

18. Which of the following is the solution for the given equation?

$$\frac{x^2 + x - 30}{x - 5} = 11$$

 a. $x = -6$
 b. All real numbers.
 c. $x = 16$
 d. $x = 5$
 e. There is no solution.

19. Mom's car drove 72 miles in 90 minutes. How fast did she drive in feet per second?
 a. 0.8 feet per second
 b. 48.9 feet per second
 c. 0.009 feet per second
 d. 70.4 feet per second
 e. 21.3 feet per second

20. Solve $V = lwh$ for h.
   a. $lwV = h$
   b. $h = \frac{V}{lw}$
   c. $h = \frac{Vl}{w}$
   d. $h = \frac{Vw}{l}$
   e. $h = \frac{Vl}{w}$

21. What is the domain for the function $y = \sqrt{x}$?
   a. All real numbers
   b. $x \geq 0$
   c. $x > 0$
   d. $y \geq 0$
   e. $x < 0$

22. If Sarah reads at an average rate of 21 pages in four nights, how long will it take her to read 140 pages?
   a. 6 nights
   b. 26 nights
   c. 8 nights
   d. 27 nights
   e. 21 nights

23. The phone bill is calculated each month using the equation $c = 50g + 75$. The cost of the phone bill per month is represented by c, and g represents the gigabytes of data used that month. Identify and interpret the slope of this equation.
   a. 75 dollars per day
   b. 75 gigabytes per day
   c. 50 dollars per day
   d. 50 dollars per gigabyte
   e. The slope cannot be determined

24. What is the function that forms an equivalent graph to $y = \cos(x)$?
   a. $y = \tan(x)$
   b. $y = \csc(x)$
   c. $y = \sin\left(x + \frac{\pi}{2}\right)$
   d. $y = \sin\left(x - \frac{\pi}{2}\right)$
   e. $y = \tan\left(x + \frac{\pi}{2}\right)$

25. If $\sqrt{1 + x} = 4$, what is x?
   a. 10
   b. 15
   c. 20
   d. 25
   e. 36

26. What is the inverse of the function $f(x) = 3x - 5$?
   a. $f^{-1}(x) = \frac{x}{3} + 5$
   b. $f^{-1}(x) = \frac{5x}{3}$
   c. $f^{-1}(x) = 3x + 5$
   d. $f^{-1}(x) = \frac{x+5}{3}$
   e. $f^{-1}(x) = \frac{x}{3} - 5$

27. What are the zeros of $f(x) = x^2 + 4$?
   a. $x = -4$
   b. $x = \pm 2i$
   c. $x = \pm 2$
   d. $x = \pm 4i$
   e. $x = 2, 4$

28. Twenty is 40% of what number?
   a. 60
   b. 8
   c. 200
   d. 70
   e. 50

29. What is the simplified form of the expression $1.2 * 10^{12} \div 3.0 * 10^8$?
   a. $0.4 * 10^4$
   b. $4.0 * 10^4$
   c. $4.0 * 10^3$
   d. $3.6 * 10^{20}$
   e. $4.0 * 10^2$

30. You measure the width of your door to be 36 inches. The true width of the door is 35.75 inches. What is the relative error in your measurement?
   a. 0.7%
   b. 0.007%
   c. 0.99%
   d. 0.1%
   e. 7.0%

31. What is the y-intercept for $y = x^2 + 3x - 4$?
   a. $y = 1$
   b. $y = -4$
   c. $y = 3$
   d. $y = 4$
   e. $y = -3$

32. Is the following function even, odd, neither, or both?

$$y = \frac{1}{2}x^4 + 2x^2 - 6$$

a. Even
b. Odd
c. Neither
d. Both
e. Even for all negative x-values and odd for all positive x-values

33. Which equation is not a function?
a. $y = |x|$
b. $y = \sqrt{x}$
c. $x = 3$
d. $y = 4$
e. $y = 3x$

34. How could the following function be rewritten to identify the zeros?

$$y = 3x^3 + 3x^2 - 18x$$

a. $y = 3x(x + 3)(x - 2)$
b. $y = x(x - 2)(x + 3)$
c. $y = 3x(x - 3)(x + 2)$
d. $y = (x + 3)(x - 2)$
e. $y = 3x(x + 3)(x + 2)$

35. A six-sided die is rolled. What is the probability that the roll is 1 or 2?
a. $\frac{1}{6}$
b. $\frac{1}{4}$
c. $\frac{1}{3}$
d. $\frac{1}{2}$
e. $\frac{1}{36}$

36. A line passes through the origin and through the point (-3, 4). What is the slope of the line?
a. $-\frac{4}{3}$
b. $-\frac{3}{4}$
c. $\frac{4}{3}$
d. $\frac{3}{4}$
e. $\frac{1}{3}$

37. What type of function is modeled by the values in the following table?

x	f(x)
1	2
2	4
3	8
4	16
5	32

a. Linear
b. Exponential
c. Quadratic
d. Cubic
e. Logarithmic

38. An investment of $2,000 is made into an account with an annual interest rate of 5%, compounded continuously. What is the total value of the investment after eight years?

a. $4,707
b. $3,000
c. $2,983.65
d. $10,919.63
e. $1,977.61

39. A ball is drawn at random from a ball pit containing 8 red balls, 7 yellow balls, 6 green balls, and 5 purple balls. What's the probability that the ball drawn is yellow?

a. $^{1}/_{26}$
b. $^{19}/_{26}$
c. $^{14}/_{26}$
d. 1
e. $^{7}/_{26}$

40. Two cards are drawn from a shuffled deck of 52 cards. What's the probability that both cards are Kings if the first card isn't replaced after it's drawn?

a. $^{1}/_{169}$
b. $^{1}/_{221}$
c. $^{1}/_{13}$
d. $^{4}/_{13}$
e. $^{1}/_{104}$

41. What's the probability of rolling a 6 at least once in two rolls of a die?

a. $^1/_3$
b. $^1/_{36}$
c. $^1/_6$
d. $^1/_{12}$
e. $^{11}/_{36}$

42. Given the set $A = \{1, 2, 3, 4, 5, 6, 7, 8, 9, 10\}$ and $B = \{1, 2, 3, 4, 5\}$,what is $A - (A \cap B)$?

a. $\{6, 7, 8, 9, 10\}$
b. $\{1, 2, 3, 4, 5\}$
c. $\{1, 2, 3, 4, 5, 6, 7, 8, 9, 10\}$
d. $\emptyset$
e. $\{-1, -2, -3, -4, -5\}$

43. An equilateral triangle has a perimeter of 18 feet. If a square whose sides have the same length as one side of the triangle is built, what will be the area of the square?

a. 6 square feet
b. 36 square feet
c. 256 square feet
d. 1000 square feet
e. 324 square feet

44. In a group of 20 men, the median weight is 180 pounds and the range is 30 pounds. If each man gains 10 pounds, which of the following would be true?

a. The median weight will increase, and the range will remain the same.
b. The median weight and range will both remain the same.
c. The median weight will stay the same, and the range will increase.
d. The median weight and range will both increase.
e. The median weight will increase, and the range will decrease.

45. For the following similar triangles, what are the values of x and y (rounded to one decimal place)?

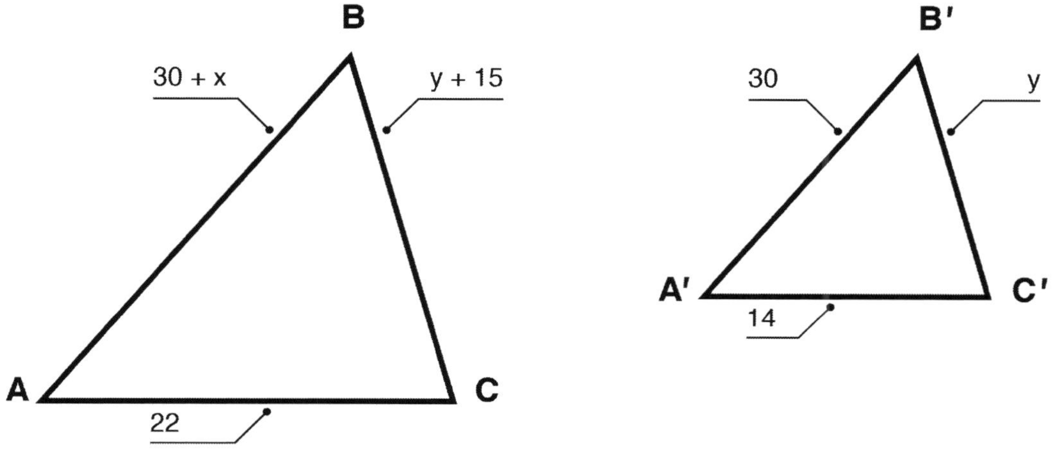

a. $x = 16.5, y = 25.1$
b. $x = 19.5, y = 24.1$
c. $x = 17.1, y = 26.3$
d. $x = 26.3, y = 17.1$
e. $x = 24.1, y = 19.5$

46. On Monday, Robert mopped the floor in 4 hours. On Tuesday, he did it in 3 hours. If on Monday, his average rate of mopping was p sq. ft. per hour, what was his average rate on Tuesday?

a. $\frac{4}{3}p$ sq. ft. per hour

b. $\frac{3}{4}p$ sq. ft. per hour

c. $\frac{5}{4}p$ sq. ft. per hour

d. $p + 1$ sq. ft. per hour

e. $\frac{1}{3}p$ sq. ft. per hour

47. Which of the following inequalities is equivalent to $3 - \frac{1}{2}x \geq 2$?

a. $x \geq 2$
b. $x \leq 2$
c. $x \geq 1$
d. $x \leq 1$
e. $x \leq -2$

48. For which of the following are $x = 4$ and $x = -4$ solutions?

a. $x^2 + 16 = 0$
b. $x^2 + 4x - 4 = 0$
c. $x^2 - 2x - 2 = 0$
d. $x^2 - \text{x} - 16 = 0$
e. $x^2 - 16 = 0$

49. What are the center and radius of a circle with equation $4x^2 + 4y^2 - 16x - 24y + 51 = 0$?
   a. Center (3, 2) and radius ½
   b. Center (2, 3) and radius ½
   c. Center (3, 2) and radius ¼
   d. Center (2, 3) and radius ¼
   e. Center (2, 2) and radius ¼

50. If the ordered pair $(-3, -4)$ is reflected over the x-axis, what's the new ordered pair?
   a. $(-3, -4)$
   b. $(3, -4)$
   c. $(3, 4)$
   d. $(-3, 4)$
   e. $(-4, -3)$

51. If the volume of a sphere is 288π cubic meters, what are the radius and surface area of the same sphere?
   a. Radius is 6 meters and surface area is 144π square meters
   b. Radius is 36 meters and surface area is 144π square meters
   c. Radius is 6 meters and surface area is 12π square meters
   d. Radius is 36 meters and surface area is 12π square meters
   e. Radius 12 meters and surface area 144π square meters

52. The triangle shown below is a right triangle. What's the value of x?

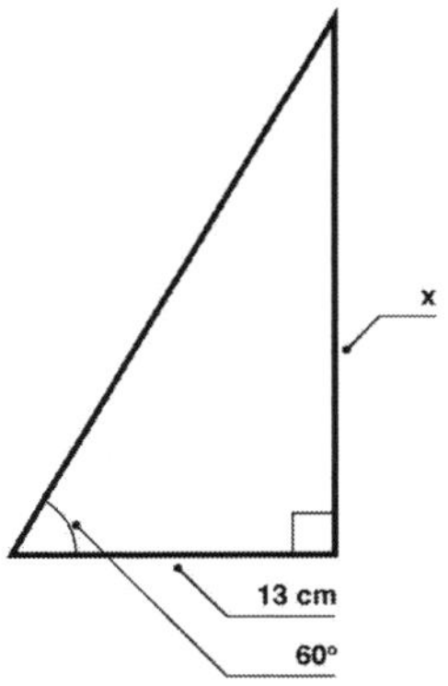

   a. $x = 1.73$
   b. $x = 0.57$
   c. $x = 13$
   d. $x = 14.73$
   e. $x = 22.52$

53. Ten students take a test. Five students get a 50. Four students get a 70. If the average score is 55, what was the last student's score?

a. 20
b. 40
c. 50
d. 60
e. 62

54. A sample data set contains the following values: 1, 3, 5, 7. What's the standard deviation of the set?

a. 2.58
b. 4
c. 6.23
d. 1.1
e. 0.25

55. A company invests $50,000 in a building where they can produce saws. If the cost of producing one saw is $40, then which function expresses the amount of money the company pays? The variable *y* is the money paid and *x* is the number of saws produced.

a. $y = 50{,}000x + 40$
b. $y + 40 = x - 50{,}000$
c. $y = 40x - 50{,}000$
d. $y = 40x + 50{,}000$
e. $y = 4x + 50{,}000$

56. A pair of dice is thrown, and the sum of the two scores is calculated. What's the expected value of the roll?

a. 5
b. 6
c. 7
d. 8
e. 9

57. A line passes through the point (1, 2) and crosses the *y*-axis at *y* = 1. Which of the following is an equation for this line?

a. $y = 2x$
b. $y = x + 1$
c. $x + y = 1$
d. $y = \frac{x}{2} - 2$
e. $y = x - 1$

58. $x^4 - 16$ can be simplified to which of the following?

a. $(x^2 - 4)(x^2 + 4)$
b. $(x^2 + 4)(x^2 + 4)$
c. $(x^2 - 4)(x^2 - 4)$
d. $(x^2 - 2)(x^2 + 4)$
e. $(x^2 - 2)(x^2 + 2)$

59. $(4x^2y^4)^{\frac{3}{2}}$ can be simplified to which of the following?

a. $8x^3y^6$

b. $4x^{\frac{5}{2}}y$

c. $4xy$

d. $32x^{\frac{7}{2}}y^{\frac{11}{2}}$

e. x^3y^6

60. A ball is thrown from the top of a high hill, so that the height of the ball as a function of time is $h(t) = -16t^2 + 4t + 6$, in feet. What is the maximum height of the ball in feet?

a. 6
b. 6.25
c. 6.5
d. 6.75
e. 6.8

Answer Explanations for Practice Test #3

1. D: This problem can be solved using basic arithmetic. Xavier starts with 20 apples, then gives his sister half, so 20 divided by 2.

$$\frac{20}{2} = 10$$

He then gives his neighbor 6, so 6 is subtracted from 10.

$$10 - 6 = 4$$

Lastly, he uses ¾ of his apples to make an apple pie, so to find remaining apples, the first step is to subtract ¾ from one and then multiply the difference by 4.

$$\left(1 - \frac{3}{4}\right) \times 4 = ?$$

$$\left(\frac{4}{4} - \frac{3}{4}\right) \times 4 = ?$$

$$\left(\frac{1}{4}\right) \times 4 = 1$$

2. C: The product of two irrational numbers can be rational or irrational. Sometimes, the irrational parts of the two numbers cancel each other out, leaving a rational number. For example, $\sqrt{2} \times \sqrt{2} = 2$ because the roots cancel each other out. Technically, the product of two irrational numbers can be complex because complex numbers can have either the real or imaginary part (in this case, the imaginary part) equal zero and still be considered a complex number. However, Choice *D* is incorrect because the product of two irrational numbers is not an imaginary number so saying the product is complex *and* imaginary is incorrect.

3. B: The car is traveling at a speed of five meters per second. On the interval from one to three seconds, the position changes by ten meters. By making this change in position over time into a rate, the speed becomes ten meters in two seconds or five meters in one second.

4. C: The number negative four is classified as a real number because it exists and is not imaginary. It is rational because it does not have a decimal that never ends. It is an integer because it does not have a fractional component. The next classification would be whole numbers, for which negative four does not qualify because it is negative. Choices *D* and *E* are wrong because -4 is not considered an irrational number because it does not have a never-ending decimal component.

5. B: Since $850 is the price *after* a 20% discount, $850 represents 80% of the original price. To determine the original price, set up a proportion with the ratio of the sale price (850) to original price (unknown) equal to the ratio of sale percentage (where x represents the unknown original price):

$$\frac{850}{x} = \frac{80}{100}$$

To solve a proportion, cross multiply the numerators and denominators and set the products equal to each other: $(850)(100) = (80)(x)$. Multiplying each side results in the equation $85{,}000 = 80x$.

To solve for x, divide both sides by 80: $\frac{85{,}000}{80} = \frac{80x}{80}$, resulting in $x = 1062.5$. Remember that x represents the original price. Subtracting the sale price from the original price ($\$1062.50 - \$850$) indicates that Frank saved \$212.50.

6. D: \$0.45

List the givens.

$$Store\ coffee = \$1.23/lbs$$

$$Local\ roaster\ coffee = \$1.98/1.5\ lbs$$

Calculate the cost for 5 lbs of store brand.

$$\frac{\$1.23}{1\ lbs} \times 5\ lbs = \$6.15$$

Calculate the cost for 5 lbs of the local roaster.

$$\frac{\$1.98}{1.5\ lbs} \times 5\ lbs = \$6.60$$

Subtract to find the difference in price for 5 lbs.

$$\begin{array}{r} \$6.60 \\ -\$6.15 \\ \hline \$0.45 \end{array}$$

7. C: The first step in solving this problem is expressing the result in fraction form. Separate this problem first by solving the division operation of the last two fractions. When dividing one fraction by another, invert or flip the second fraction and then multiply the numerator and denominator.

$$\frac{7}{10} \times \frac{2}{1} = \frac{14}{10}$$

Next, multiply the first fraction with this value:

$$\frac{3}{5} \times \frac{14}{10} = \frac{42}{50}$$

In this instance, you can find the decimal form by converting the fraction into $\frac{x}{100}$, where x is the number from which the final decimal is found. Multiply both the numerator and denominator by 2 to get the fraction as an expression of $\frac{x}{100}$.

$$\frac{42}{50} \times \frac{2}{2} = \frac{84}{100}$$

In decimal form, this would be expressed as 0.84.

8. E: To find the average of a set of values, add the values together and then divide by the total number of values. In this case, include the unknown value of what Dwayne needs to score on his next test, in order to solve it.

$$\frac{78 + 92 + 83 + 97 + x}{5} = 90$$

Add the unknown value to the new average total, which is 5. Then multiply each side by 5 to simplify the equation, resulting in:

$$78 + 92 + 83 + 97 + x = 450$$

$$350 + x = 450$$

$$x = 100$$

Dwayne would need to get a perfect score of 100 in order to get an average of at least 90.

Test this answer by substituting back into the original formula:

$$\frac{78 + 92 + 83 + 97 + 100}{5} = 90$$

9. D: 1, 2, 3, 4, 6, 12. A given number divides evenly by each of its factors to produce an integer (no decimals). The number 5, 7, 8, 9, 10, 11 (and their opposites) do not divide evenly into 12. Therefore, these numbers are not factors.

10. E: The augmented matrix that represents the system of equations has dimensions 4×3 because there are three equations with three unknowns. The coefficients of the variables make up the first three columns, and the last column is made up of the numbers to the right of the equals sign. This system can be solved by reducing the matrix to row-echelon form, where the last column gives the solution for the unknown variables.

11. B: There are two zeros for the function: $x = 0, -2$.

The zeros can be found several ways, but this particular equation can be factored into:

$$f(x) = x(x^2 + 4x + 4) = x(x + 2)(x + 2)$$

By setting each factor equal to zero and solving for x, there are two solutions. On a graph these zeros can be seen where the line crosses the x-axis.

12. D: This problem involves a composition function, where one function is plugged into the other function. In this case, the $f(x)$ function is plugged into the $g(x)$ function for each x-value. The composition equation becomes:

$$g(f(x)) = 2^3 - 3(2^2) - 2(2) + 6$$

Simplifying the equation gives the answer:

$$g(f(x)) = 8 - 3(4) - 2(2) + 6 = 8 - 12 - 4 + 6 = -2$$

13. D: This system of equations involves one quadratic function and one linear function, as seen from the degree of each equation. One way to solve this is through substitution. Solving for y in the second equation yields $y = x + 2$. Plugging this equation in for the y of the quadratic equation yields:

$$x^2 - 2x + x + 2 = 8$$

Simplifying the equation, it becomes:

$$x^2 - x + 2 = 8$$

Setting this equal to zero and factoring, it becomes:

$$x^2 - x - 6 = 0 = (x - 3)(x + 2)$$

Solving these two factors for x gives the zeros $x = 3, -2$. To find the y-value for the point, each number can be plugged in to either original equation. Solving each one for y yields the points $(3, 5)$ and $(-2, 0)$.

14. D: The expression is simplified by collecting like terms. Terms with the same variable and exponent are like terms, and their coefficients can be added.

15. A: Finding the product means distributing one polynomial onto the other. Each term in the first must be multiplied by each term in the second. Then, like terms can be collected. Multiplying the factors yields the expression:

$$20x^3 + 4x^2 + 24x - 40x^2 - 8x - 48$$

Collecting like terms means adding the x^2 terms and adding the x terms. The final answer after simplifying the expression is:

$$20x^3 - 36x^2 + 16x - 48$$

16. E: Finding the zeros for a function by factoring is done by setting the equation equal to zero, then completely factoring. Since there was a common x for each term in the provided equation, that would be factored out first. Then the quadratic that was left could be factored into two binomials, which are $(x + 1)(x - 4)$. Setting each factor equal to zero and solving for x yields three zeros.

17. A: 13 nurses

Using the given information of 1 nurse to 25 patients and 325 patients, set up an equation to solve for number of nurses (N):

$$\frac{N}{325} = \frac{1}{25}$$

Multiply both sides by 325 to get N by itself on one side.

$$\frac{N}{1} = \frac{325}{25} = 13\ nurses$$

18. E: The equation can be solved by factoring the numerator into $(x + 6)(x - 5)$.

Since that same factor exists on top and bottom, that factor $(x - 5)$ cancels.

This leaves the equation $x + 6 = 11$.

Solving the equation gives the answer $x = 5$. When this value is plugged into the equation, it yields a zero in the denominator of the fraction. Since this is undefined, there is no solution.

19. D: This problem can be solved by using unit conversion. The initial units are miles per minute. The final units need to be feet per second. Converting miles to feet uses the equivalence statement 1 mile equals 5,280 feet. Converting minutes to seconds uses the equivalence statement 1 minute equals 60 seconds. Setting up the ratios to convert the units is shown in the following equation:

$$\frac{72 \text{ mi}}{90 \text{ min}} \times \frac{1 \text{ min}}{60 \text{ s}} \times \frac{5280 \text{ ft}}{1 \text{ mi}} = 70.4\frac{\text{ft}}{\text{s}}$$

The initial units cancel out, and the new units are left.

20. B: The formula can be manipulated by dividing both the length, *l*, and the width, *w*, on both sides. The length and width will cancel on the right, leaving height by itself.

21. B: The domain is all possible input values, or *x*-values. For this equation, the domain is every number greater than or equal to zero. There are no negative numbers in the domain because taking the square root of a negative number results in an imaginary number.

22. D: This problem can be solved by setting up a proportion involving the given information and the unknown value. The proportion is:

$$\frac{21\, pages}{4\, nights} = \frac{140\, pages}{x\, nights}$$

Solving the proportion by cross-multiplying, the equation becomes $21x = 4 \times 140$, where $x = 26.67$. Since it is not an exact number of nights, the answer is rounded up to 27 nights. Twenty-six nights would not give Sarah enough time.

23. D: The slope from this equation is 50, and it is interpreted as the cost per gigabyte used. Since the *g*-value represents number of gigabytes and the equation is set equal to the cost in dollars, the slope relates these two values. For every gigabyte used on the phone, the bill goes up 50 dollars.

24. C: Graphing the function $y = \cos(x)$ shows that the curve starts at $(0, 1)$, has an amplitude of 2, and a period of 2π. This same curve can be constructed using the sine graph, by shifting the graph to the left $\frac{\pi}{2}$ units. This equation is in the form:

$$y = \sin(x + \frac{\pi}{2})$$

25. B: Start by squaring both sides to get $1 + x = 16$. Then subtract 1 from both sides to get $x = 15$.

26. A: This inverse of a function is found by switching the x and y in the equation and solving for y. In the given equation, solving for y is done by adding 5 to both sides, then dividing both sides by 3. This answer can be checked on the graph by verifying the lines are reflected over $y = x$.

27. B: The zeros of this function can be found by using the quadratic formula:

$$x = \frac{-b \pm \sqrt{b^2 - 4ac}}{2a}$$

Identifying a, b, and c can also be done from the equation because it is in standard form. The formula becomes:

$$x = \frac{0 \pm \sqrt{0^2 - 4(1)(4)}}{2(1)} = \frac{\sqrt{-16}}{2}$$

Since there is a negative underneath the radical, the answer is a complex number:

$$x = \pm 2i$$

28. E: Setting up a proportion is the easiest way to represent this situation. The proportion becomes $\frac{20}{x} = \frac{40}{100}$, where cross-multiplication can be used to solve for x. The answer can also be found by observing the two fractions as equivalent, knowing that twenty is half of forty, and fifty is half of one-hundred.

29. C: Division with scientific notation can be solved by grouping the first terms together and grouping the tens together. The first terms can be divided, and the tens terms can be simplified using the rules for exponents. The initial expression becomes $0.4 * 10^4$. This is not in scientific notation because the first number is not between 1 and 10. Shifting the decimal and subtracting one from the exponent yields $4.0 * 10^3$.

30. A: The relative error can be found by finding the absolute error and making it a percent of the true value. The absolute error is $36 - 35.75 = 0.25$. This error is then divided by 36—the true value—to find 0.7%.

31. B: The y-intercept of an equation is found where the x-value is zero. Plugging zero into the equation for x allows the first two terms to cancel out, leaving -4.

32. A: The equation is *even* because:

$$f(-x) = f(x)$$

Plugging in a negative value will result in the same answer as when plugging in the positive of that same value. The function:

$$f(-2) = \frac{1}{2}(-2)^4 + 2(-2)^2 - 6 = 8 + 8 - 6 = 10$$

yields the same value as:

$$f(2) = \frac{1}{2}(2)^4 + 2(2)^2 - 6 = 8 + 8 - 6 = 10$$

33. C: The equation $x = 3$ is not a function because it does not pass the vertical line test. This test is made from the definition of a function, where each x-value must be mapped to one, and only one, y-value. This equation is a vertical line, so the x-value of 3 is mapped with an infinite number of y-values.

34. A: The function can be factored to identify the zeros. First, the term $3x$ is factored out to the front because each term contains $3x$. Then, the quadratic is factored into $(x + 3)(x - 2)$.

35. C: A die has an equal chance for each outcome. Since it has six sides, each outcome has a probability of $\frac{1}{6}$. The chance of a 1 or a 2 is therefore $\frac{1}{6} + \frac{1}{6} = \frac{1}{3}$.

36. A: The slope is given by:

$$m = \frac{y_2 - y_1}{x_2 - x_1} = \frac{0 - 4}{0 - (-3)} = -\frac{4}{3}$$

37. B: The table shows values that are increasing exponentially. The differences between the inputs are the same, while the differences in the outputs are changing by a factor of 2. The values in the table can be modeled by the equation $f(x) = 2^x$.

38. C: The formula for continually compounded interest is $A = Pe^{rt}$. Plugging in the given values to find the total amount in the account yields the equation:

$$A = 2000e^{0.05*8} = 2983.65$$

39. E: The sample space is made up of $8 + 7 + 6 + 5 = 26$ balls. The probability of pulling each individual ball is $^1/_{26}$. Since there are 7 yellow balls, the probability of pulling a yellow ball is $^7/_{26}$.

40. B: For the first card drawn, the probability of a King being pulled is $\frac{4}{52}$. Since this card isn't replaced, if a King is drawn first the probability of a King being drawn second is $\frac{3}{51}$. The probability of a King being drawn in both the first and second draw is the product of the two probabilities:

$$\frac{4}{52} \times \frac{3}{51} = \frac{12}{2652}$$

This fraction, when divided by 12, equals $\frac{1}{221}$.

41. E: The addition rule is necessary to determine the probability because a 6 can be rolled on either roll of the die. The rule used is:

$$P(A \text{ or } B) = P(A) + P(B) - P(A \text{ and } B)$$

The probability of a 6 being individually rolled is $\frac{1}{6}$ and the probability of a 6 being rolled twice is:

$$\frac{1}{6} \times \frac{1}{6} = \frac{1}{36}$$

Therefore, the probability that a 6 is rolled at least once is:

$$\frac{1}{6}+\frac{1}{6}-\frac{1}{36}=\frac{11}{36}$$

42. A: $(A \cap B)$ is equal to the intersection of the two sets A and B, which is $\{1,2,3,4,5\}$. $A-(A \cap B)$ is equal to the elements of A that are *not* included in the set $(A \cap B)$. Therefore:

$$A-(A \cap B)=\{6,7,8,9,10\}$$

43. B: An equilateral triangle has three sides of equal length, so if the total perimeter is 18 feet, each side must be 6 feet long. A square with sides of 6 feet will have an area of $6^2=36$ square feet.

44. A: If each man gains 10 pounds, every original data point will increase by 10 pounds. Therefore, the man with the original median will still have the median value, but that value will increase by 10.

The smallest value and largest value will also increase by 10 and, therefore, the difference between the two won't change. The range does not change in value and, thus, remains the same.

45. C: Because the triangles are similar, the lengths of the corresponding sides are proportional. Therefore:

$$\frac{30+x}{30}=\frac{22}{14}=\frac{y+15}{y}$$

This results in the equation $14(30+x)=22\times 30$ which, when solved, gives $x=17.1$.

The proportion also results in the equation $14(y+15)=22y$ which, when solved, gives $y=26.3$.

46. A: Robert accomplished his task on Tuesday in ¾ the time compared to Monday. He must have worked 4/3 as fast.

47. B: To simplify this inequality, subtract 3 from both sides to get $-\frac{1}{2}x \geq -1$. Then, multiply both sides by -2 (remembering this flips the direction of the inequality) to get $x \leq 2$.

48. E: There are two ways to approach this problem. Each value can be substituted into each equation. Choice *A* can be eliminated, since:

$$4^2+16=32$$

Choice *B* can be eliminated, since:

$$4^2+4\cdot 4-4=28$$

Choice *C* can be eliminated, since:

$$4^2-2\cdot 4-2=6$$

But, plugging in either value into x^2-16 gives:

$$(\pm 4)^2-16=16-16=0$$

48. E: There are two ways to approach this problem. Each value can be substituted into each equation. Choice *A* can be eliminated, since:

$$4^2 + 16 = 32$$

Choice *B* can be eliminated, since:

$$4^2 + 4 \cdot 4 - 4 = 28$$

Choice *C* can be eliminated, since:

$$4^2 - 2 \cdot 4 - 2 = 6$$

But, plugging in either value into $x^2 - 16$ gives:

$$(\pm 4)^2 - 16 = 16 - 16 = 0$$

49. B: The technique of completing the square must be used to change:

$$4x^2 + 4y^2 - 16x - 24y + 51 = 0$$

into the standard equation of a circle. First, the constant must be moved to the right-hand side of the equals sign, and each term must be divided by the coefficient of the x^2 term (which is 4). The x and y terms must be grouped together to obtain:

$$x^2 - 4x + y^2 - 6y = -\frac{51}{4}$$

Next, the process of completing the square must be completed for each variable. This gives:

$$(x^2 - 4x + 4) + (y^2 - 6y + 9) = -\frac{51}{4} + 4 + 9$$

The equation can be written as:

$$(x - 2)^2 + (y - 3)^2 = \frac{1}{4}$$

Therefore, the center of the circle is (2, 3) and the radius is:

$$\sqrt{1/4} = 1/2$$

50. D: When an ordered pair is reflected over an axis, the sign of one of the coordinates must change. When it's reflected over the x-axis, the sign of the y-coordinate must change. The x-value remains the same. Therefore, the new ordered pair is $(-3, 4)$.

51. A: Because the volume of the given sphere is 288π cubic meters, this means $4/3\,\pi r^3 = 288\pi$. This equation is solved for r to obtain a radius of 6 meters. The formula for the surface area of a sphere is $4\pi r^2$, so if $r = 6$ in this formula, the surface area is 144π square meters.

52. E: SOHCAHTOA is used to find the missing side length. Because the angle and adjacent side are known, $\tan 60 = \frac{x}{13}$. Making sure to evaluate tangent with an argument in degrees, this equation gives:

$$x = 13 \tan 60 = 13 \times \sqrt{3} = 22.52$$

53. A: Let the unknown score be x. The average will be:

$$\frac{5 \times 50 + 4 \times 70 + x}{10} = \frac{530 + x}{10} = 55$$

Multiply both sides by 10 to get $530 + x = 550$, or $x = 20$.

54. A: First, the sample mean must be calculated:

$$\bar{x} = \frac{1}{4}(1 + 3 + 5 + 7) = 4$$

The standard deviation of the data set is $s = \sqrt{\frac{\sum(x-\bar{x})^2}{n-1}}$, and $n = 4$ represents the number of data points. Therefore:

$$s = \sqrt{\frac{1}{3}[(1-4)^2 + (3-4)^2 + (5-4)^2 + (7-4)^2]} = \sqrt{\frac{1}{3}(9 + 1 + 1 + 9)} = 2.58$$

55. D: For manufacturing costs, there is a linear relationship between the cost to the company and the number produced, with a y-intercept given by the base cost of acquiring the means of production, and a slope given by the cost to produce one unit. In this case, that base cost is \$50,000, while the cost per unit is \$40. So, $y = 40x + 50{,}000$.

56. C: The expected value is equal to the total sum of each product of individual score and probability. There are 36 possible rolls.

The probability of rolling a 2 is $\frac{1}{36}$.

The probability of rolling a 3 is $\frac{2}{36}$.

The probability of rolling a 4 is $\frac{3}{36}$.

The probability of rolling a 5 is $\frac{4}{36}$.

The probability of rolling a 6 is $\frac{5}{36}$.

The probability of rolling a 7 is $\frac{6}{36}$.

The probability of rolling an 8 is $\frac{5}{36}$.

The probability of rolling a 9 is $\frac{4}{36}$.

The probability of rolling a 10 is $\frac{3}{36}$.

The probability of rolling an 11 is $\frac{2}{36}$.

Finally, the probability of rolling a 12 is $\frac{1}{36}$.

Each possible outcome is multiplied by the probability of it occurring. Like this:

$$2 \times \frac{1}{36} = a$$

$$3 \times \frac{2}{36} = b$$

$$4 \times \frac{3}{36} = c$$

And so forth.

Then all of those results are added together:

$$a + b + c \ldots = expected\ value$$

In this case, it equals 7.

57. B: From the slope-intercept form, $y = mx + b$, it is known that b is the y-intercept, which is 1. Compute the slope as $\frac{2-1}{1-0} = 1$, so the equation should be $y = x + 1$.

58. A: This has the form $t^2 - y^2$, with $t = x^2$ and $y = 4$.

It's also known that:

$$t^2 - y^2 = (t + y)(t - y)$$

and substituting the values for t and y into the right-hand side gives $(x^2 - 4)(x^2 + 4)$.

59. A: Simplify this to:

$$(4x^2y^4)^{\frac{3}{2}} = 4^{\frac{3}{2}}(x^2)^{\frac{3}{2}}(y^4)^{\frac{3}{2}}$$

Now:

$$4^{\frac{3}{2}} = (\sqrt{4})^3 = 2^3 = 8$$

For the other, recall that the exponents must be multiplied, so this yields:

$$8x^{2\cdot\frac{3}{2}}y^{4\cdot\frac{3}{2}} = 8x^3y^6$$

60. B: The independent variable's coordinate at the vertex of a parabola (which is the highest point, when the coefficient of the squared independent variable is negative) is given by $x = -\frac{b}{2a}$. Substitute and solve for *x* to get:

$$x = -\frac{4}{2(-16)} = \frac{1}{8}$$

Using this value of *x*, the maximum height of the ball (*y*), can be calculated.

Substituting *x* into the equation yields:

$$h(t) = -16\frac{1}{8}^2 + 4\frac{1}{8} + 6 = 6.25$$

Greetings!

First, we would like to give a huge "thank you" for choosing us and this study guide for your ACT exam. We hope that it will lead you to success on this exam and for your years to come.

Our team has tried to make your preparations as thorough as possible by covering all of the topics you should be expected to know. In addition, our writers attempted to create practice questions identical to what you will see on the day of your actual test. We have also included many test-taking strategies to help you learn the material, maintain the knowledge, and take the test with confidence.

We strive for excellence in our products, and if you have any comments or concerns over the quality of something in this study guide, please send us an email so that we may improve.

As you continue forward in life, we would like to remain alongside you with other books and study guides in our library, such as;

ACCUPLACER: amazon.com/dp/1628457724

SAT: amazon.com/dp/1628457856

We are continually producing and updating study guides in several different subjects. If you are looking for something in particular, all of our products are available on Amazon. You may also send us an email!

Sincerely,
APEX Test Prep
info@apexprep.com

FREE

Free Study Tips DVD

In addition to the tips and content in this guide, we have created a FREE DVD with helpful study tips to further assist your exam preparation. **This FREE Study Tips DVD provides you with top-notch tips to conquer your exam and reach your goals.**

Our simple request in exchange for the strategy-packed DVD is that you email us your feedback about our study guide. We would love to hear what you thought about the guide, and we welcome any and all feedback—positive, negative, or neutral. It is our #1 goal to provide you with top quality products and customer service.

To receive your **FREE Study Tips DVD**, email freedvd@apexprep.com. Please put "FREE DVD" in the subject line and put the following in the email:

a. The name of the study guide you purchased.

b. Your rating of the study guide on a scale of 1-5, with 5 being the highest score.

c. Any thoughts or feedback about your study guide.

d. Your first and last name and your mailing address, so we know where to send your free DVD!

Thank you!

Made in the USA
Columbia, SC
05 August 2021

42987386R00096